6 95/MC

Arithmetical Disabilities
In Cerebral Palsied
Children

Arithmetical Disabilities In Cerebral Palsied Children

Programmed Instruction—A Remedial Approach

By

SIMON H. HASKELL, Ph.D.

*Tutor and Lecturer in the Education of
Physically Handicapped Children
Department of Child Development
Institute of Education
University of London*

With a Foreword by

Peter Mittler, Ph.D.

*Director
Hester Adrian Research Centre
University of Manchester
Manchester, England*

CHARLES C THOMAS · PUBLISHER
Springfield · Illinois · U.S.A.

Published and Distributed Throughout the World by
CHARLES C THOMAS • PUBLISHER
BANNERSTONE HOUSE
301-327 East Lawrence Avenue, Springfield, Illinois, U.S.A.

This book is protected by copyright. No part of it may be reproduced in any manner without written permission from the publisher.

© *1973, by* CHARLES C THOMAS • PUBLISHER
ISBN 0-398-02537-1
Library of Congress Catalog Card Number: 72-75920

With THOMAS BOOKS *careful attention is given to all details of manufacturing and design. It is the Publisher's desire to present books that are satisfactory as to their physical qualities and artistic possibilities and appropriate for their particular use.* THOMAS BOOKS *will be true to those laws of quality that assure a good name and good will.*

Printed in the United States of America
Q-1

To

JAMES LUMSDEN

good friend of handicapped children

FOREWORD

Although a great deal of progress has been made in providing schools for cerebral palsied children, very little research has so far been devoted to methods of teaching them. In fact, we know more about their intellectual and educational difficulties than about techniques of overcoming or compensating for them.

This book, by Dr. Simon Haskell, tutor and lecturer in the education of handicapped children at the University of London, represents one of the few attempts to redress the balance. Its relevance is by no means confined to teachers of cerebral palsied children, because it helps the reader to think of arithmetical competence not merely as a global undifferentiated ability but as a series of separately identifiable skills. These skills depend for their effective functioning on a complex system of sensory, perceptual, motor, intellectual and motivational factors. Any or all of these systems can be affected in a child with cerebral palsy, and it is therefore axiomatic that effective remedial teaching should be based on comprehensive assessment leading to the design of a teaching programme which is so far as possible suited to the specific needs of the individual.

The use of programmed learning techniques in the education of handicapped children helps the teacher to think more systematically about the content of teaching and how the material can be presented to a child with complex perceptual or learning problems. But a teaching machine is only as good as its programmes and unfortunately commercial interests and our fascination with machines have outstripped our capacity to provide properly prepared and educationally relevant programmes. The research reported in this book used a commercial machine and a commercial programme to teach arithmetical skills. I hope that teachers will be stimulated by this research to develop a personal interest in programmed learning and educational technology, to create a demand for courses in these subjects in colleges and universities, and finally to develop and evaluate their own programmes.

PETER MITTLER

INTRODUCTION

Whilst the medical and social provision for cerebral palsied children has greatly increased during the last two decades, in Great Britain little research has been undertaken to study the effects of cerebral palsy on learning in such children.

One of the neglected areas of research has been a study of the arithmetic skills and knowledge of such children. The learning of arithmetic by normal children would be impeded by limited intellectual abilities, inappropriate teaching, impoverished environment and emotional disorder. Some cerebral palsied children are likely to experience additional difficulty in learning arithmetic because of impairment in perceptual and visuomotor abilities as well as on account of their lack of mobility and manipulative skills. These are known to impoverish experience and to contribute to emotional difficulties. Failure to recognize these factors often leads to inappropriate teaching.

PROGRAMMED INSTRUCTION

Professor Skinner has shown that programmed instruction can help to overcome some of the difficulties experienced by normal children. Children are able to work individually and at their own pace. By skillful arrangement of the subject matter into small, well-graded steps, and by immediate and frequent information about the correctness of his responses, learning is facilitated and the child can proceed from very simple to complex steps. Skinner claims that in such a learning climate far fewer emotional demands are made on the child. The machine in which the programme is housed never tires, has unlimited patience, and cannot scold the child if it makes mistakes. Moreover, the teacher is relieved of such routine and monotonous chores as marking exercise books, although this advantage may be destroyed by extra time in supervising the loading and management of the machine.

Also, according to Skinner, individual differences are not of crucial importance. In fact, programmed instruction can capitalise on them. No longer are differences in intelligence an obstacle to learning, because the dull student has only to work longer than the bright to master difficult concepts (Fry, 1960).

Skinner (1962) declares that "modern children do not learn arithmetic quickly or well." He regards the conventional methods of teaching arithmetic to young children as clumsy, boring, uneconomical and most likely to produce emotional disturbance. Two main criticisms emerge from his analysis of the inadequacies of the conventional methods. First, he sees serious limitations in having children learn in order to avoid punishment. By punishment he means not only physical measures, such as caning, the use of the birch and other devices, but also ridicule, threat and criticism from teachers and classmates. Second, he deplores the failure to reinforce learning adequately in arithmetic. Skinner estimates that out of a possible 25,000 contingencies of reinforcement necessary for the learning of mathematics in the first four years of school, the average child receives only a meagre few thousand. Conventional teaching methods create delays of up to twenty-four hours before a child is told whether his answers were correct. This delay is inevitable when teachers have to examine many exercise books at a time and often return them a day later.

The purpose of the present study was to review possible factors causing difficulties experienced by cerebral palsied children in learning arithmetic and to undertake an exploratory study to see how far programmed instruction could overcome some of these difficulties. This would include:

1. An assessment of arithmetic attainment. This can be measured in three different ways.
    a. Calculating skills
       An understanding of four basic rules and an ability to compute accurately and speedily.

b. Reasoning
Verbal reasoning ability as measured on such items of standardized IQ tests as Wechsler Intelligence Scale for Children (WISC) subtests. This includes an appreciation of money values, measurement of heights, weights and distances.
c. Practical performance
The practical and social applications of the use of money such as giving correct change, or estimating measurements with reasonable precision.

Whilst the wider aspects of arithmetic attainment were considered important, attention was focused only on children's "calculating skills."
2. A review of the relevant literature relating to various factors likely to influence calculating skills. Information may be gained from studies relating to three main groups of children: normal, mentally retarded, and "brain-injured," including cerebral palsied children.
3. An exploration and evaluation of appropriateness of using programmed instruction to teach cerebral palsied children calculating skills.

At a recent meeting the following definition of cerebral palsy was accepted:

> [Cerebral palsy is] a disorder of movement and posture due to a defect or lesion of the immature brain . . . it is usual to exclude from cerebral palsy those disorders of posture and movement which are (1) of short duration, (2) due to progressive disease and (3) due solely to mental deficiency (Bax, 1964).

With some modification, the classification suggested by the Little Club Clinics in Developmental Medicine (MacKeith et al., 1959) has been adopted.

SPASTIC CEREBRAL PALSY

SPASTIC { Hemiplegia
Diplegia
Double hemiplegia

ATHETOID { Dystonic cerebral palsy
Choreo-athetoid cerebral palsy

OTHERS { Mixed forms of cerebral palsy
Ataxic cerebral palsy
Atonic diplegia

This study is only concerned with the three main types (spastic, athetoid and ataxic). These types make up about 95 percent of cases of cerebral palsy. The other types (dystonic, atonic diplegia and mixed cerebral palsies) are rare and did not occur in the children studied. The atonic ("floppy"), dystonic (disordered muscle tone) and mixed (in which the predominant feature appears with other clinical symptoms) types of cerebral palsies make up about 5 percent of all cases.

ACKNOWLEDGMENTS

It is a pleasure to acknowledge my indebtedness to Mr. J. J. Q. Fox for his invaluable help and encouragement. My thanks are also due to Dr. Marcel Kinsbourne, Dr. Franz Morgenstern and Miss Elizabeth Anderson for their valued criticisms aimed at clarifying the text.

I am grateful to Mr. James Loring for his continued interest and support and to the Spastics Society for a research grant to undertake this investigation. I am indebted to Miss Joanna Wormald who typed the manuscript with such expert care, and to Miss Melinda Nelson for help in preparing the index.

My thanks are particularly due to the Headteachers, staff, and children whose friendliness and cooperation made this research such a happy and profitable experience.

I would also like to thank the editors of the *Journal of Mental Subnormality* and the *Journal of Special Education* for permission to use material from previous articles listed below:

1. Haskell, Simon H.: Programmed instruction and the mentally retarded.

    In Gunzburg, H. C. (Ed.): *The Application of Research to the Education and Training of the Severely Subnormal Child.* Monograph Supplement, *Journal of Mental Subnormality*, 1966.
2. Haskell, Simon H.: Impairment of arithmetic skills in cerebral palsied children and a programmed remedial approach. *Journal of Special Education*, I (No. 4):Summer, 1967.

CONTENTS

	Page
Foreword — Peter Mittler	vii
Introduction	ix
Acknowledgments	xiii

PART ONE
Chapter
I. REVIEW OF THE LITERATURE 5

PART TWO
II. EXPERIMENT 39
III. RESULTS 48
IV. DISCUSSION 65

PART THREE
V. IMPLICATIONS FOR TEACHERS 79

Summary 92
Bibliography 96
Appendix 103
Index 105

Arithmetical Disabilities In Cerebral Palsied Children

PART ONE

CHAPTER I

REVIEW OF THE LITERATURE

THE REVIEW OF THE literature consists of five sections. In the first section the factors affecting the arithmetic attainment of non-cerebral palsied children will be discussed. The literature relating to the general learning difficulties encountered by cerebral palsied children is next reviewed, and is followed by an analysis of the possible factors likely to affect their arithmetic attainment specifically. Finally in the remaining two sections the evidence is examined which suggests programmed instruction could play an important part in the teaching of arithmetic, and especially to cerebral palsied children.

The five sections discussed are listed below:

1. Factors affecting arithmetic attainment in non-cerebral palsied children.
2. Learning difficulties of cerebral palsied children.
3. Cerebral palsied children and arithmetic.
4. Programmed instruction and arithmetic.
5. Programmed instruction for cerebral palsied children.

FACTORS AFFECTING ARITHMETIC ATTAINMENT IN NON-CEREBRAL PALSIED CHILDREN

Interest in the "quantitative expression" of young normal children dates back several decades, and has given rise to an extensive literature. Dutton (1964) reviewed the research findings under six headings: arithmetic readiness, motivation, thinking and concept development, transfer of learning, evaluation of procedures and teachers' understanding of arithmetic.

Though normal children have been studied in the full variety of their activities, namely, from the developmental and differential standpoints, there has been a notable lack of coordination in research strategy. Buswell (1958) drew attention to the lack of a conceptual framework in most recent studies. Biggs (1959) stressed the value of discovering how children come to acquire basic mathematical concepts, rather than attempt to measure what children know.

We shall examine the literature relevant to the following sources of difficulty with arithmetic in normal children: intellectual, emotional, neurological, preschool experiences, and school experiences.

Intellectual Factors

In some studies reported in the *Sixty-Second Year Book* (1963), it was noted that children in the lowest ability range did poorly compared with the brighter children. The differences in arithmetic attainment were attributed to individual differences in intelligence.

Bruner (1959), Dienes (1959), Lovell (1961) and others stress that children's capacity to deal with new experiences and develop mathematical concepts vary and are related to differences in intelligence at any one age level. Piaget (1952, 1953[a], 1953[b]) repeatedly drew attention to the effects of differences in intelligence on learning ability. Whilst he maintains that all children must progress through the same sequential stages in mathematical ideas, diversities in development arise because of variations in intelligence, maturation and experience.

Several studies show that mentally retarded children are inferior to normal children in arithmetic attainments (Cruickshank, 1948; Capobianco, 1956; Kirk, 1962). Kirk (1962) reviewing the studies of arithmetic achievement in mentally retarded children describes four characteristic weaknesses: (a) poor ability at mental arithmetic tasks, especially when presented verbally, (b) excessive dependence on "counting on fingers," (c) careless work habits and (d) poor understanding of mathematical concepts. Some studies show that mentally retarded children have adequate mechanical computational skills in accordance with their mental age, but are deficient in arithmetical reasoning tasks. Moreover, with skilled teaching some children show considerable improvement in calculating but fail to achieve comparable understanding of mathematical concepts. Cruickshank in three studies showed that mentally retarded children seemed unable to select the relevant facts and procedures in arriving at a correct solution, and were markedly inferior to normal children in arithmetic "work habits." They also employed "primitive" procedures in simple computational tasks, were significantly slower than the children of normal intelligence in understanding abstract concepts and were extremely careless in their work habits.

In the first study Cruickshank (1948) tested the assumption that mentally retarded children, unlike normal children, fail to solve arithmetic problems because they could not exclude extraneous facts while attempting a correct solution. His study confirmed that mentally retarded children were unable to select relevant data because of inferior language development and comprehension. Furthermore they were deficient in reasoning ability.

The second investigation by Cruickshank (1948[a]), carried out with the same children, set out to study whether mentally retarded and normal children alike understood and solved arithmetic problems effectively. The results indicated that normal children were significantly superior in identifying as well as solving arithmetic problems. It was noted that mentally retarded children had greater success in solving arithmetic problems requiring a concrete rather than an abstract solution. He observed a marked tendency on the part of mentally retarded children to change their mode of operation while engaged in problem solving.

Finally Cruickshank (1948[b]) examined the arithmetic attainment of these two groups of children. The Buswell-John Diagnostic Chart for fundamentals of arithmetic was administered and the results showed a consistent inferiority in the performance of the mentally retarded children. These children made characteristically poorer responses in simple computational tasks. Often they used primitive procedures such as finger counting, marked their papers when adding, subtracting, dividing and multiplying, thus making numerous mistakes. They had special difficulty in using the zero digit, guessed frequently and were often careless in their habits.

Emotional Factors

Burt (1937) observed that anxious children appeared to be better at reading than at arithmetic. Berakat (1951) found that character qualities correlate with mathematical attainments to the extent of 0.30 to 0.35. This was an overall correlation derived from a study of three hundred grammar school pupils, and in the case of individual pupils encountered, emotional factors were on occasion more important than intellectual ones. He noted that emotional instability seemed to correlate most highly with inaccuracy of computation and lack of industry with inefficiency in mathematical reasoning. Other workers (cited in Lynn, 1957) confirmed that arithmetic disability is related to temperamental characteristics. Lynn (1957) mentions a series of studies in which it was reported that emotionally disturbed children demonstrated a disparity between reading and arithmetic attainment. The speculation is offered that anxious children read to escape into fantasy and this helps in the resolution of conflicts, or allays their anxieties. It is often difficult to establish whether poor performance in arithmetic and repeated failure in a specific learning situation causes disturbed behaviour. Contrary evidence is offered by Gregory (1965) who has shown that unsettledness or maladjustment as measured on the Bristol Social Adjustment Guide had been a "contributory cause" to reading failure among primary school children in a West Berkshire village. Most of the cases reported are clinical studies suggesting that anxiety is positively associated with poor arithmetic attainment. However, since it has been well established that the acquisition of complex skills is impaired by anxiety, a possible basis for a relationship between anxiety and effective mathematical thinking can be attempted.

Biggs (1959) also reviews the few studies showing a relationship between emotional factors and poor arithmetic attainment.

He considers that fear of arithmetic, particularly working with numbers, is especially marked in some maladjusted children.

However, there are encouraging accounts of intelligent dyscalculic children responding to appropriate remedial teaching. But the danger of some children with initial difficulties with "number work" becoming severely disturbed by unsympathetic and unskilled teaching is emphasised by Biggs. He also suggested that children exposed early in life to negative attitudes to arithmetic by parents and various social groups are predisposed to identify themselves with these attitudes.

Feldhusen and Klausmeier (1962) reported a study in which the effects of anxiety upon children's scholastic attainments, including arithmetic, was noted (arithmetic, reading and language). One hundred and twenty children who were studied were divided equally into three groups and differentiated on the basis of their WISC-IQ ratings, as set out in the table following:

TABLE I

DISTRIBUTION OF GROUPS AND IQ RANGES

Groups	Number	IQ Range
1	40	56- 81
2	40	90-110
3	40	120-146
Total	120	

The children were tested on the California Achievement Battery and the Children's Manifest Anxiety Scale. The authors found that the children in the lowest IQ group were significantly more anxious than the other two and that there was a negative correlation between anxiety and IQ for each of the three groups. In the lowest IQ group anxiety and poor arithmetic attainment was significantly correlated (at the 1 percent level) and the girls were more anxious than boys.

Other studies showing a disparity in arithmetic attainment due to sex differences fail to take account of any cultural determinants. There are indications that girls fulfil certain social expectations in that they under-achieve in arithmetic as compared with boys. Sometimes there are overt and covert implications that "cleverness" in arithmetic leads to a loss of "femininity" in girls.

Some writers claim that children's arithmetic attainment could yield significant "subtle and diagnostic clues to emotional and intellectual development, and that improvement in emotionally disturbed children is often reflected in progress in arithmetic" (Newman et al., 1964). In the psychoanalytic literature numerous case studies are mentioned in which computational skills and mathematical concepts are reported to be affected by unconscious processes. Various psychopathological states like obsessions, phobias, "slips," "learning blocks," etc. involving arithmetic are related to unconscious psychological conflicts. Several of these reports have been listed by Newman et al. (1964) but the evidence from these studies is mostly anecdotal and many of the factors mentioned are not necessarily contributory ones.

There is, however, evidence in the studies reviewed previously that emotional factors may contribute significantly to poor attainment in arithmetic, and that of these, temperamental characteristics, particularly anxiety, and emotional disturbances caused by the unsympathetic attitudes of teachers towards a child's initial failures are of major importance.

Neurological Factors

Some workers felt that arithmetic difficulties were caused by neurological disorders. The inspiration for these studies was drawn from Gerstmann (1924) who drew attention to a syndrome (Gerstmann's syndrome) of cerebral cortical disorder. These clinical abnormalities which he described consisted of an inability to "recognise, indicate on request, name, or choose with open eyes, individual fingers either on their own hands or the hands of others." Gerstmann suggested that a deficiency in number operations (acalculia) was associated with this disorder. Subsequently attempts to relate arithmetic (and writing) difficulties in mentally retarded and brain injured subjects to neurological abnormalities have been made by several people. Unfortunately the earlier studies were anecdotal accounts of the conditions and rash conclusions were drawn on the basis of a few clinical cases. Strauss and Werner (1938) and Werner and Carrison (1942) examined the possibility that a deficiency in the finger schema contributed to arithmetic difficulties in children who exhibited certain clinical abnormalities observed earlier by Gerstmann. The authors administered "Finger Schema" tests to small groups of children who were selected on the basis of arithmetic attainment. On the basis of four cases it was concluded that an impairment in the finger schema in mentally retarded children contributed to special arithmetic disability.

Benton et al. (1951) compared right-left discrimination, finger localisation and arithmetic ability in two groups of children of normal and subnormal intelligence. They found no evidence to support the hypothesis that disordered finger localisation was positively associated with poor arithmetic achievement. The authors were very critical of the Strauss and Werner investigation on account of the slender evidence offered in support of their speculations.

Kinsbourne and Warrington (1963) reported that a group of thirteen children who had been referred for backwardness in reading and writing could be differentiated on the basis of the discrepancy between the verbal and performance scores of the WISC. The group of children who had significantly higher verbal scores on the WISC had associated difficulties of finger differentiation and order together with impaired mechanical arithmetic ability. Mathews and Folk (1964) noted that with mentally retarded subjects success in finger localisation was positively correlated with written arithmetic.

While these studies suggest some relationship between neurological disorders and poor arithmetic attainment they add little of substance to our knowledge of the causes of difficulty in arithmetic in non-cerebral palsied children since they are occasionally contradictory and generally inconclusive.

Preschool Experiences

Children attending primary school for the first time appear to learn arithmetic more effectively if they have had appropriate preschool experiences in social and play situations (Dutton, 1964). If provided with natural or carefully planned opportunities to explore their environment, children learn to discover by sensori-motor activities the properties of objects around them. Thus by the conscious exploitation and manipulation of such play materials as bricks, stacking cubes, matching rods, etc. in various play situations, the foundations of mathematical concepts are carefully laid down. These physical experiences prepare the child for more systematic instruction at school.

Brownell (1941) carried out a major study to assess the arithmetic readiness of primary grade children in America. He indicated that children enter school equipped with varying levels of information about the subject. In the same investigation, Brownell referred to the studies of Woody, who showed that children's attainment, even in rote counting, depended largely upon "someone assuming the duty of teaching arithmetic to children." There was clear evidence also that children's preschool arithmetic achievement was related to such factors as the intellectual socioeconomic home background and motivation of parents.

Piaget has drawn attention to the fact that young children are incapable of forming ideas of objects unless they have abundant opportunities to perform physical actions with them. But while stressing the importance of experiences which relate to the formation of number concepts and number vocabulary and to the early basis of simple spatial knowledge, he does warn educators that concrete materials do not in themselves teach children concepts but that it is the manipulation of them which makes evident the significance of the operations.

Churchill (1961) also drew attention to the importance of the "operational character" of mathematical learning in young children since they depend almost exclusively upon bodily operations rather than verbal behaviour.

School Experiences

Teachers' Attitude to Arithmetic and Understanding of Concepts
The crucial role of the teacher in helping children form

mathematical concepts by providing appropriate physical and concrete experiences is stressed by educators and psychologists alike. As mentioned earlier children may fail to understand mathematical concepts because of the unfavourable attitudes of teachers. Also, some teachers who have an inadequate understanding of mathematical concepts themselves are unable to instruct children in an intelligible way. Glennon (1948), studying the reasons why children lack understanding in arithmetic, mentioned among six causes the failure on the part of teachers to make arithmetic meaningful. He observed that the average "seventh-grade" pupil had acquired only 12.5 percent of the understanding basis to the computational processes taught in grades one through six. Moreover, teachers who taught new concepts prematurely to children have either received faulty instruction at school or training college. Such teachers were unable themselves to understand the basic concepts underlying mathematics. Dutton (1961) carried out a study to assess student teachers' understanding of arithmetic concepts. Two classes of fifty-five "elementary" school teachers were examined in this respect and it was noted that a substantial number relied on traditional methods and mechanical procedures learnt during their school days.

Cheney (1961) carried out a more systematic review examining the extent to which teachers' failure to present arithmetic meaningfully stemmed from their own inadequate understanding of the subject. He noted that some teachers experience difficulties in understanding specific concepts, and that many primary grade teachers have poor skills in teaching concepts beyond the simplest stages. Cheney drew attention to the need for the teacher to understand concepts rather than to train children in mechanical skills. Lovell (1961) also mentions that the acquisition of mathematical concepts by children is dependent on adequate understanding by the teacher and the "climate of opinion in which the child is reared."

Appropriate Intervention in Teaching

Churchill (1961) drew attention to the teacher's role in

securing the interest of the child by making the learning situation meaningful. The teacher was required to create an environment that aroused curiosity, provided challenging experiences and stimulated children to experiment. Piaget also stressed that the skilled teacher is one who can maintain rapport with the children by knowing when to intervene, and what experiences are relevant to the acquisition of specific skills. He postulated two invariant sequences of development through which the child must proceed before he is capable of understanding mathematical concepts. These he described as the "preoperational" stage in which the child cannot understand even the simplest of quantitative relationships, making judgments only by using crude perceptual approximations, and the "intuitive" stage in which the child can, for example, appreciate the idea of conservation intellectually, although he finds this difficult to equate with the apparent inconsistency of his perceptions. It is not until the third or "operational" stage that the child is ready to study logical number with real understanding.

Thus according to Piaget it would be meaningless for a teacher to attempt to teach a child the concept of number before he has grasped the idea of conservation, and if teachers are to enable children to understand concepts rather than merely to develop mechanical skills they must be aware of the sequences in which these concepts develop.

Teaching Aids

Teachers are expected to take account of the differing levels of ability, the prior experience of children, their emotional status and rates at which they learn. These make high demands on insight, skill and knowledge and it is sometimes argued, especially by Skinner (1962), that most teachers are ill-equipped to discharge this function.

Teachers who are familiar with the variety of structural apparatus devised for teaching arithmetic (Williams, 1961) and know little about the rationale behind these systems, often rely on formal blackboard instruction. The disadvantages of the latter approach for very young children are apparent. On the other hand, the indiscriminate use of concrete aids in instructing young children is inappropriate and advised against by Van Engen and Gibbs (1960). They warned about the dangers in the misuse of teaching aids and stressed the value of providing the correct physical and concrete experiences in arithmetic for young children.

Lovell (1961) stressed that the teacher has an important role in providing suitable teaching apparatus for seven to eleven year olds. Lunzer (1961) suggests that there are manifest advantages in a teacher progressing through the logical stages with a child learning a new concept, rather than waiting for a spontaneous discovery.

Computational Practice

Schonell and Schonell (1957), in a survey of the factors influencing arithmetic attainment, point out that pupils who display adequate knowledge of processes and effective reasoning ability may yet consistently fail to get the right answers because of insufficient properly planned practice of the four rules. While emphasising the undesirability of practice without understanding, they note that understanding without practice yields similarly poor results and refer to Meddleton (1954) who carried out an experimental investigation into the systematic teaching of number combinations in arithmetic. Meddleton demonstrated that the systematic use of scientifically compiled sheets of number combinations gave results in arithmetic attainment superior to the usual unplanned drills of the classroom. Although the control groups in the experiment were given more practice in addition and multiplication they tended to neglect subtraction and division, while the experimental class, because it used properly planned material, was given adequate practice in the combinations in all four processes. Schonell and Schonell (1957) cite this study as providing "the most carefully checked statistical evidence to date of the marked advantage to teachers of following a systematic programme with properly planned material in order to develop accuracy and speed in the number combinations."

Absence from School

Schonell and Schonell (1957) also list absence from school, whether intermittent or prolonged, as one of the most important causes of lack of ability in arithmetic. They point out that arithmetic is more susceptible to the influence of absences than any other school subject, since whereas a child who has mastered the mechanics of reading can practice at home, this does not apply to arithmetic where new steps are so numerous and systematic, and regular practice with graded examples so important to automatize past steps and to consolidate new ones. Teachers should be aware of the feelings of dismay which absence from arithmetic may create in certain children.

Summary

While much of the literature relating to the factors which affect arithmetic attainment in non-cerebral palsied children has been shown to be tentative and inconclusive, a number of factors of indisputable significance have emerged, and these are summarized below:

1. Differences in intelligence, at any one age level, determine to a considerable extent a pupil's success in arithmetic.
2. Mentally retarded children are inferior to normal children in overall arithmetical attainments, and demonstrate:
   a. a poor understanding of mathematical concepts and a deficiency in arithmetical reasoning tasks.
   b. an inability to select relevant data often because of inferior language development and comprehension.
   c. the use of primitive procedures in simple computational tasks, especially an excessive dependence on finger counting.
   d. careless work habits.
3. Emotional factors often contribute significantly to poor attainment in arithmetic.
4. Temperamental characteristics, particularly the presence of anxiety, can be correlated with poor arithmetic attainment.
5. Unsympathetic attitudes on the part of teachers may

contribute to an emotional disturbance which inhibits progress in arithmetic.

6. Preschool experiences relating to the formation of number concepts, number vocabulary and simple spatial knowledge, largely determine arithmetic readiness.

7. Appropriate experiences of this kind involve the child's having the opportunity for considerable sensorimotor activity, including the active manipulation of objects.

8. The teacher's ability to make arithmetic meaningful depends to a large extent on his own understanding of arithmetic concepts, an understanding which is frequently lacking.

9. The teacher must be aware of the child's level of development so that he can supply experiences appropriate for him at that stage.

10. Frequently little use is made of structural apparatus and reliance is placed on formal blackboard instruction, or the available special apparatus may not be used properly.

11. There is often a lack of enough carefully planned computational practice in schools.

12. Absence from school is likely to have a worse effect on attainment in arithmetic than in other subjects.

LEARNING DIFFICULTIES OF CEREBRAL PALSIED CHILDREN

It is widely known that injury to the brain sustained in infancy is attended by an increase in motor (Dunsdon, 1952) and intellectual disorders. See Table II.

These may result in generalised or highly specific learning disabilities. Among the intellectual disabilities most commonly identified in cerebral palsied children are perceptual disorders of various types, and it is the impairment of visual perceptual skills which has received the greatest attention from investigators. A number of studies have therefore been made of perceptual responses of these children usually through standardized psychological tests. Most of the more detailed findings are inconclusive, although the observations of Cruickshank, Bice and Waller that

TABLE II

SOME STUDIES SHOWING DISTRIBUTION OF INTELLIGENCE OF
CEREBRAL PALSIED CHILDREN IN BRITAIN AND AMERICA
AND COMPARISON WITH THE NORMAL POPULATION

IQ Range	Normal Population—Asher and Schonell	Birmingham Asher and Schonell (N = 354)	Dundee Henderson (N = 223)	Selected Areas in Britain Dunsdon (N = 916)	New Jersey Cruickshank and Raus (N = 1,000)	California From Dunsdon (N = 113)
	Percentage			Incidence		
130 and over	1	0.6	8.2	0.5	1.6	0.8
100-129	24	3.4	8.1	2.0	5.3	7.8
90 109	46	20.1	16.1	6.2	21.6	25.4
70- 89	24	26.8	25.1	15.5	22.7	31.0
50- 69	3	22.9	21.1	17.2	20.4	26.3
Below 50		22.3	27.4	23.6	28.4	8.7

cerebral palsied children, especially spastics, perform poorly compared with normals on visual and visuomotor perceptual tasks has been generally confirmed (Nielsen, 1966).

The most significant early studies of these problems were, however, made by Strauss and Lehtinen (1947) who were largely influenced by the Gestalt psychologists and who devised special educational methods for "brain injured" children based on their clinical observations, using materials not unlike the Montessori apparatus. Strauss and Lehtinen defined a brain injured child as "a child, who, before, during or after birth has received an injury to or suffered an infection of the brain. As a result of such organic impairment, defects of the neuromotor system may be present or absent; however, such a child may show disturbances in perception, thinking and emotional behaviour, either separately or in combination."

The concept of brain injury has often been used very loosely; Rutter (1967) suggests that it is most usefully employed as a general term which would include "a number of different syndromes, or patterns of difficulty which are the result of some kind of impaired functioning of the brain. This use of the term would certainly include the cerebral palsied child, but would also include many more children with less severe and more subtle handicaps such as difficulties in coordination, perception and speech." He thus divides the syndromes resulting from brain damage into two broad groups, firstly those involving definite abnormalities of function, where the two main conditions involved are cerebral palsy and epilepsy, and secondly those in which there are limits or delays in the development of normal function.

Strauss and Lehtinen's definition of the term "brain-injured" would cover both groups and would thus include children with cerebral palsy. From their studies of such children they concluded that perceptual disturbances contributed largely to their generally poor educational attainments in reading, writing and arithmetic. They drew attention to the fact and to the educational implications of certain children's inability to distinguish between figure and background, to their distractability (Goldstein's forced responsiveness to stimuli), to their lack of success in "integrating elements into comprehensive configurations," to their tendency to "perseverate" and to the signs of "rigidity" which they showed in discriminating tasks. These children made bizarre and inappropriate responses to such tasks as sorting objects and ordering them in a logical sequence. In their emotional behaviour they were noticed to react in an overwhelming ("catastrophic") manner to ordinary experiences or to mild provocation, and frequently their behavior could be described as "erratic," uncoordinated, uncontrolled, disinhibited and socially unacceptable.

Today it is recognised more clearly that such behavior and such intellectual abnormalities do not occur in all "brain injured" children and that, on the other hand "normal" children in infant or junior schools may exhibit, though usually in somewhat milder forms, some of these patterns and difficulties.

Strauss and Lehtinen offered practical suggestions for the teaching of the mentally retarded brain injured (exogenous) child and emphasise that it should differ from that provided for the non-brain-injured mentally retarded (endogenous) child. They maintain that the exogenous pupils could be taught with advantage only in smaller classes (not exceeding 12). Specialised methods and aids for teaching arithmetic, reading and writing are advocated and the authors give hints about reducing distractability in children. The reduction of visual and auditory stimuli in the classroom was effected by allowing children to sit facing blank walls, removing pictures, masking extraneous material and other such expedients. Teachers even ceased wearing bracelets, earrings, dangling necklaces and other ornaments. Aids to facilitate increased understanding of arithmetic such as counting boxes, numbers and dots on rotating wheels and other structural apparatus are described.

Children with visual and auditory perceptual disorders were helped to overcome their weaknesses and confusions in reading by special remedial techniques. These included the use of graded apparatus to increase perceptual discrimination. Writing skills were improved by specific training in copying through onionskin on tracing paper or on clay pans.

Hopkins et al. (1954) carried out a general survey of the factors relating to cerebral palsy. They examined the medical, psychological and educational records of 1,505 cerebral palsied children. Of this group more detailed psychological information was obtained on ninety-nine pupils attending a special school. Their findings indicated that the median IQ was 79, and 35 percent had IQs of below 70, whilst only 4 percent had IQs of 110 or above. All the children were educationally retarded. Unfortunately this study was only a general summary and the possibilities of remedial education were not explored.

Cruickshank et al. (1957) studied the largest number of cerebral palsied children and his is the most comprehensive account to date. While it was concluded that the cerebral palsied groups performed less well than normals on five out of six tests, no general perceptual impairment was evident in such children. Wedell (1960) who confirmed Cruickshank's findings further noted that differences exist not only between athetoids and spastics but also between left and right hemiplegics on visuoperceptual and visuomotor tasks.

Cruickshank's subjects included 211 spastics, 114 athetoids and 110 normal children aged six to sixteen years. A number of tests were administered to assess the figure ground perceptual ability in these children. As this is one of the most important studies in the literature it is especially disappointing that they failed to control for intelligence, and the unwarranted assumption that the higher IQs of the control group did not affect the results is unjustified. Moreover the failure to examine all the children on a standardized IQ test was a serious omission, and the weakness of sampling might explain the lack of consistency in other findings.

Kephart (1966) believes that a basic problem of the brain injured child is an interference with the ability to generalise (that is to organise or integrate his perceptions) and therefore stresses that teaching should be directed to development of generalisation. He points out that generalisation is achieved firstly through the acquisition of a "datum" and secondly through the elaboration of the datum by the acquisition of a large number of similar but not identical experiences. He believed that while the normal child introduces such variations himself, the brain injured child may not do this but will repeatedly perform the activity in exactly the same way. Teaching should therefore involve the continual introduction of large numbers of variations and the presentation of the same material in different ways (that is, through more than one sense avenue) at the same point in time. Kephart believes that the combined sources of information will present the child with the required similarity and that thus his ability to make abstractions can be guided.

Gallagher (1960) conducted a valuable experiment into the effects of tutoring "brain injured mentally retarded" children for an hour a day for approximately 120 to 150 days. Forty-two children between the ages of seven years four months and thirteen years nine months, with IQs from 33 to 63 (Stanford Binet) were divided equally into two groups according to "mental development." All the chidren attended a school for handicapped pupils in Illinois. One group was "tutored" for the first two years but not the third, whilst the other group was "untutored" for the first two years but "tutored" in the third. Comprehensive data on each child obtained prior to the tutoring programme included their verbal and nonverbal IQs as well as information about their perceptual, language and mathematical development. Soon after the experiment commenced, personality ratings were obtained, whilst the battery of tests were administered at intervals of nine, twenty-one and thirty-three months. The four tutors were flexible in their instructional methods, but focused their attention on such aspects as the perceptual, language, memory, conceptualisation reasoning and simple numerical processes.

Gallagher found that all children benefited from tutoring, but regressed when it was discontinued; and that there were positive transfer effects of tutoring to nontutored areas of development, these gains being maximised in the younger child.

One of the most comprehensive investigations of various aspects of cerebral palsied children in Britain and elsewhere was carried out by Dunsdon (1952). From a central register of 3,700 children (newborn to 16 yr of age) she studied the case records of 1,044 children drawn from selected areas in England and Wales. Except for a few instances she made individual assessments of the abilities, aptitudes and attainments of 916 children. Some of her findings are well summarised by Reid (1963).

Results

1. Only one out of every ten cerebral palsied children could sit by the age of one year and only four in ten had accomplished this by the age of two years. In general, the higher the level of intelligence, the earlier the child could sit.
2. Verbal abilities of the children with cerebral palsy are far below the nonhandicapped, with 79 percent defective in speech.
3. Visual impairments were found in one third of the cases, 85 percent of the children had a hearing loss of more than ten percent, and 15 percent a loss of more than 20 percent.
4. Fifty percent of the children studied were considered emotionally unstable.
5. The level of progress in school was related to the child's mental capacity. However, in many instances the contrary was observed.

Another valuable British investigation was that by Schonell (1956) who looked at the relationship between intelligence and motor disorders in cerebral palsied children. She also offered useful hints about the educational programme for such children. Schonell compared the IQs (Stanford Binet), sensory defects, involvement of limbs, speech impairments and reading attainments of 340 children suffering from various types of cerebral palsy. In contrast with Dunsdon's (1952) findings, she observed that the mean IQ of spastics and athetoids was similar (67.5) and that the ataxics and mixed types showed no differences in intelligence (62.3 to 62.4). But her finding that the general tendency for the more severely handicapped to be in the inferior IQ range is in keeping with Dunsdon's (1952) observations. Schonell further noted that the average and above average cerebral palsied child was retarded in reading, possibly because of impoverished experiences.

CEREBRAL PALSIED CHILDREN AND ARITHMETIC

The effects of damage to the brain on academic attainments, particularly arithmetic in cerebral palsied children, has attracted relatively little attention. In a few studies the performance of such children has been observed to be notably inferior to that of normals. Some of these differences are partly accounted for by variations in intelligence, emotional traits, impoverished preschool and school experiences, frequent absence from school as well as ill-chosen instructional techniques. However, it is claimed that cerebral palsied children encounter additional and outstanding difficulties in arithmetic and particularly in understanding mathematical concepts. These difficulties are said to arise because of disturbance both in peripheral and central abilities and the selective impairment of these abilities is not easily identified by intelligence tests. It is not yet widely recognised that while tests of general intelligence have modest success in predicting educational attainment with normals, their value for cerebral palsied children is extremely limited. The overall IQ score is a misleading guide, particularly when the nature of the specific intellectual loss is not identified. For example, such children could sustain greater or minor losses of perceptual or visuomotor skills, or demonstrate impaired reasoning or memory abilities. Phillips and White (1964) argue that if the predictive validity of standardised intelligence tests were measured by the "much more relevant criterion of learning in schools, then the evidence presented suggests that the mental operations sampled by the Stanford Binet and other similar tests have such limited relevance to the acquisition of basic educational skills by children with very early motor defects that they should not form the principal basis for individual decisions."

Standardised tests narrow in coverage and high in specificity would yield more accurate measures of narrow well-defined abilities.

Peripheral Disorders and Arithmetic Attainment

Hebb (1949) thought visual perception important in learning in infants. He indicated that the capacity to recognise and discriminate simple forms and shapes was a gradual and complex process. Eye movements according to Hebb's speculation "contribute constantly and essentially to perceptual integration, even though they are not the whole origin of it."

A few surveys (Smith, 1963) suggest that cerebral palsied children show an abnormally high incidence of ocular defects, particularly squint (strabismus). The psychological and educational implications of such disorders, in relation to learning have been discussed by Abercrombie (1960, 1963a, 1963b, 1964). These children have difficulty recognising shapes, matching and discriminating forms, and so forth. Piaget (1952, 1953) also draws attention to the role of spatial ability in the development of the concept of numbers in young children. Abercrombie argues that as early learning is dependent upon the integrity of visual functioning, cerebral palsied children with visual disorders seem highly vulnerable. She has attempted to differentiate children with perceptual weaknesses from those with visuomotor disorders. By visuomotor she refers to skills requiring "movement under visual control."

Abercrombie et al. (1963b) gave three groups of children two simple visual tasks. The children were physically handicapped, cerebral palsied and normal, between the ages of five and sixteen. They were asked to look at seven spots successively in a row and return to the first spot on reaching the last one, and to follow a toy electric train entering and leaving a tunnel. Electrooculograph records of their horizontal saccadic and pursuit eye movements were obtained. The performance of the cerebral palsied group was the poorest and this was associated with squinting.

Central Disorders and Arithmetic Attainment

Some of the most far-reaching effects of brain injury manifest themselves in the disturbance of visuomotor skills. These are revealed in such tests as copying a diamond, or constructing a three cube pyramid or making designs with coloured blocks. The failure to do any of these relatively simple tasks is seldom known to arise because of perceptual defects. This becomes obvious when children readily reject their own bizarre efforts on being presented with another correct copy.

Various attempts to explain the cause of visuomotor impairment have been advanced. Some workers like Floyer (1955) have suggested it is due to a developmental lag, and others have suggested that it is the result of perceptual distortion.

Motor Disorder

Younger children build up knowledge of objects in space by information from bodily movements transmitted along various sensory channels. If they are restricted in their activities by motor handicaps, particularly in early life, they would become deprived of a wide range of sensorimotor experiences. According to Piaget this would arrest the normal intellectual development of children. Some cerebral palsied children are deprived of opportunities to physically explore their environment, and many suffer an abnormal reduction of preschool experiences. They are unable to manipulate various objects and toys, cannot participate in normal play activities, such as building towers, pulling and pushing equipment etc., and seem unable to develop the appropriate symbols and concepts of mathematics.

Phillips and White (1964) compared twenty-three cerebral palsied children with motor handicap appearing in early life with thirty-two physically handicapped children without congenital brain injury. The children in the groups were selected from the same school and matched as nearly as possible for age and physical handicap. The arithmetic and reading attainments of both groups were measured and compared on the Burt (rearranged) World Reading Test and the revised Southend Arithmetic Test in Mechanical Arithmetic. After allowing for age and IQ differences, the differences in attainment between the two groups remains significant.

"It is not merely that children with motor handicaps from their early months are more backward on the average than other handicapped children. More important, the multiple correlations for this group are lower. They are significantly lower for the acquisition of arithmetic skills."

Brain Injury and Arithmetic Attainment

Dunsdon (1952) suggested that particular types of lesions in cerebral palsy produced characteristic impairment in arithmetic attainment. On comparing the reading and arithmetic achievements of thirty-five cerebral palsied children on six measures, including Gestalt maturation, verbal reasoning, visual memory, visual reasoning, verbal memory and vocabulary, she concluded that "weak gestalt appreciation" as measured on the Bender Gestalt Test was commoner amongst athetoids than spastics, and that this contributed to poorer arithmetical attainment in athetoids.

The East Scotland survey (Henderson, 1961) confirmed that cerebral palsied children were backward in arithmetic. The arithmetic attainment of 153 cerebral palsied children over the age of seven years was evaluated on Burt's Four Rules and Ballard's One Minute Test in oral addition and subtraction. The results clearly demonstrated that 93.5 percent of the children studied were found to be backward, as compared with normal children.

The authors concluded that the same causes producing difficulties in arithmetic with normals, together with additional factors, impair arithmetic attainment in cerebral palsied children.

In their discussion of the causes of arithmetic difficulties in brain injured children, Strauss and Lehtinen (1947) do not clearly distinguish between the effects of perceptual and visuomotor disorders. They speculate that weaknesses in perception and abstraction contribute to poor arithmetic attainment in some cerebral palsied children.

They hypothesise that brain injured children were specially impaired in arithmetical abilities because "the brain injured organism lacks the ability possessed by the normal child to discover spontaneously the significant relationships of the number system. Accepting the thesis that a perceptual scheme of visual spatial organisation is the basis of calculation, it is reasonable to anticipate that the organism whose ability to construct such a perceptual scheme has been disturbed will be hindered in any activities which require its use." The authors develop the point that such children would have difficulty in forming number concepts because of disordered "visual spatial organisation." "So intimate is the association between the perception of numerical relationships and the ability to reason and abstract, that lack of the former has been considered diagnostic of deficiency in the latter."

Bensberg (1953) on the other hand, studying the academic achievement of two groups of institutionalised female mental defectives, with and without brain injuries, found that there were no significant differences in arithmetic attainment between the two groups.

One of the best studies of arithmetic attainment in brain injured children was that carried out by Capobianco (1956). He examined the "qualitative and quantitative" arithmetic work habits of exogenous and endogenous mentally handicapped subjects. He compared twenty-eight brain injured subjects between the ages of ten years four months and twenty-nine years nine months, with Mental Ages between six years two months and eleven years eight months, and IQ 43-78 (Stanford Binet, 1937). Capobianco was unable to confirm Lehtinen's speculations that brain injured subjects experience greater difficulties than non brain injured in arithmetic.

Distractability

Some cerebral palsied children are known to have very short attention span, and are unable to persist in a given activity for an appropriate length of time. This tendency to become distracted by extraneous stimuli is documented by several observers (Strauss and Lehtinen, 1947, Cruickshank and Raus, 1955 and Taylor, 1959). Some of these children have difficulty in distinguishing figure from background with multiple stimuli.

Several clinical descriptions of this condition have been offered. These children are reported to be restless, hyperactive and inattentive. The effect of such behaviour on visuomotor tasks is very marked and characterised by poorly integrated movements, bizarre reproductions of copied figures and forms and an irrelevant attention to details. This disorder also affects tactual-motor performance (Dolphin and Cruickshank, 1952) and the handling of objects and material that are used in mathematical teaching becomes impaired.

Emotional Factors

The effect of emotional factors upon attainment in arithmetic has already been discussed for non-cerebral palsied children and several workers were shown to confirm the relationship between certain temperamental characteristics, especially anxiety, and a lack of ability in arithmetic.

Several investigators have observed that brain injured children are at risk psychiatrically, and therefore it seems likely that emotional factors may affect attainment in arithmetic of cerebral palsied children even more markedly than in the case of non-cerebral palsied children.

Various speculations are offered as to the cause of the higher incidence of behaviour disturbance in the neurologically impaired. Most workers have drawn inferences from retrospective studies and a number have identified such factors as prematurity, leading to difficulty in handling at home and producing the characteristic attitudes in parents of "overprotection," "anxiety" or rejection of the infant (Shirley, 1939; Berkow, 1949; Howard and Morral, 1952; Drillien, 1964). Roger *et al.* (1955) and Passamanick *et al.* (1956) noted that behaviour disturbances at school age were associated with pre and perinatal complications. Knobloch et al. (1956) suggested that abnormal behaviour in the brain injured resulted from disturbed feeding relationships.

Among the few prospective studies was that by Prechtl and Dijkstra (1960) who carried out a comparative investigation of 218 full-term infants with ascertained birth complications and sixty "normal" full-term infants. In a follow up of their development it was reported that only 12 percent of the normal group showed any "disturbance" in behaviour as compared with 38 percent of the children with a history of pre or perinatal complications and 70 percent of those diagnosed as neurologically impaired during the prenatal period.

In another prospective study, Drillien (1964) followed up the progress of over three hundred premature children in Edinburgh. At about seven years of age the children were examined, among other things, for emotional and personality development on the Stott Social Adjustment Guide (Day School version). There was strong evidence that when compared with mature full-term "controls" the premature group showed greater behaviour disturbances. Also the premature children of each sex were more disturbed at school, and the boys were less stable or adjusted than the girls.

Nielsen (1966) reviewing the literature on personality studies of cerebral palsied children suggests that on the basis of the few existing studies, a tentative conclusion would be that emotional disturbances are more common among cerebral palsied than among "physically healthy" children, but that the actual personality disorders found in such children are not specific for cerebral palsy.

Summary

In conclusion it is suggested that while the factors affecting arithmetic in non-cerebral palsied children are equally applicable to the cerebral palsied group, the following additional factors specific to cerebral palsy have also to be taken into account:

1. In the case of motor handicap, a deprivation of sensorimotor experience, especially in the preschool period.
2. A higher incidence of ocular defects, especially squint,

leading to less efficiency than normals in simple tasks (for example, in computation), involving movement of the eyes.
3. Disorders of perception leading to difficulties in recognising shapes, in matching and discriminating forms, in distinguishing figure from background, and in integrating the constituent elements to form a whole.
4. Frequent disturbances in visuomotor skills, leading to a generally poorer performance on visuomotor tasks such as copying or constructing shapes than among normal children.
5. Difficulty in making generalisations affecting the child's ability to grasp mathematical concepts.
6. A greater distractability than in normal children, which will particularly affect arithmetic attainment since "one of the major determinants of success in working sums is the degree to which the pupil can keep his mind persistently on the task in hand (Schonell and Schonell, 1957).
7. A tendency to perseverate.
8. A higher incidence of emotional disturbance than in normals.

PROGRAMMED INSTRUCTION AND ARITHMETIC

Programmed instruction is a practical application of Professor Skinner's experimental analysis of behaviour in animals and man. Previously psychologists seemed preoccupied with the various stimuli that modified behaviour, but Skinner maintained that it was impractical to control and impossible to observe the many subtle stimuli operating at any one time on an individual. Disregarding theory, he studied empirically the way an individual operated on the environment, and was in turn informed by the environment of changes occurring. Skinner argued that the behaviour of an individual could be modified and shaped, by selecting carefully only the appropriate responses emitted, and positively rewarding (reinforcing) only those responses which began to approximate to the final (terminal) behaviour it was intended to teach. In subsequent writings, he elaborated on the methods whereby any required behaviour could be elicited and stabilised. Originally it was held that, for behaviour to be established, every response needed reinforcing. Further research indicated that reinforcement is also effective when it is available at fixed or at variable intervals and ratios. Another important aspect of programmed instruction was its emphasis on positive rewards (reinforcement) for learning, particularly in the classroom, as superior to any form of punishment. Though others like Pressey pioneered the idea, Skinner revolutionised educational thinking by providing a direct and simple application of operant conditioning for the teacher. He expressed regret that whilst great changes in many areas of human activity were occurring, our schools had failed to share in the exciting technological advances. Skinner remains highly critical of any teaching method in which children "learn" in order to avoid punishment. He expressed disappointment that so few students derive "a certain amount of pleasure and satisfaction from the educational process," because even the basic school subjects like reading, spelling and arithmetic are taught inadequately, and hold little meaning for the individual. This could be remedied, he argues, if the subject matter to be learnt were "programmed" or split up into small sequential steps, allowing the children to proceed logically with frequent prompts as to the correctness of his responses. Skinner dramatically demonstrated this technique with pigeons and other animals in the laboratory before advocating its use in the classroom.

The material to be learned can be arranged in two ways. One way, the linear method (also known as the "constructed response" or "extrinsic" type of programme) consists of a series of small steps through which the student progresses. Each step is presented in the form of short statements or single items called "frames," each requiring the answer to a question or the completion of a sentence, equation, etc. The sequence of steps is designed to ensure that the learner is correct in at least 95 percent of his responses. The linear type of programme is also sometimes known as the "Skinnerian" type since it is closely based on the principles of operant conditioning in that the student learns by emitting or constructing a response, he progresses gradually to complex repertoires, he receives the immediate reinforcement which Skinner thought essential and he is aided in his learning through knowledge of the results (a concept which Skinner stressed) and finally, in many programmes of this type the principle of "fading" or gradual withdrawal of stimulus support is employed.

The second main type of programme is known as the "branching" or "multiple choice" type, and was developed by Norman Crowder. Crowder describes this type of programme thus: "the student is given the material to be learned in small logical units and is tested on each unit immediately. The test result is used automatically to control the material the student sees next. If the student passes the test information he is given the next unit of information and the next question. If he fails it, the preceding unit of information is reviewed, the nature of his error is explained to him and he is retested. The test questions are multiple-choice and there is a separate set of corresponding materials for each wrong answer that is included in the multiple-choice alternative. Because it is the student's own answer which determines the material to which he will next be exposed, this has also been called "intrinsic" programming.

The programmes are usually presented by machines, though Skinner himself warned against the dangers of overemphasising the role of the machine instead of the programmed material. In fact, programmed material can be presented in book form and prove equally effective. Branching programmes often require more complex "hardware" and the material is displayed on films or cards. Another device for presenting the material is the scrambled textbook, in which the order of the frames does not coincide with those pages in the book. The linear programmes are usually housed in simpler machines that display a single frame at a time and a window to feed back confirmatory information. Most have cheat-proof devices and some are fitted with a counter to record the number of errors made.

The attempt to use programmed instruction for teaching arithmetic to normal subjects includes those made by Pressey in 1927, although his was strictly a testing device. Skinner (1954) constructed a machine for teaching arithmetic to young children and later David Zeaman devised a special purpose machine at the University of Connecticut to teach elementary arithmetic to students. Other studies (Porter, 1957; Forster and Sapon, 1958; Keislar, 1959) showed that programmed instruction facilitates understanding of arithmetic instead of rote learning. Keislar (1959) using programmed instruction to test arithmetical understanding in fourteen elementary school pupils, matched carefully for IQ, sex, reading ability and pretest scores with fourteen controls, concluded that the "brighter children" benefited more than the others from programmed instruction.

Investigations have also been carried out using programmed instruction with mentally retarded subjects but whilst there is a rapidly increasing literature on its use with normal subjects, less than two dozen studies on the mentally retarded have been reported (Stolurow, 1960, 1961; Silberman, 1962; Birnbrauer et al. 1964). Stolurow (1963) reviews fourteen studies including seven using language and five using arithmetic programmes. Some of the earlier investigators were modest in their scope and merely sought to demonstrate that mentally retarded subjects could be taught equally well by programmed instruction and always with positive results. Stolurow points out that these studies were of limited value, because of a lack of normative data based on alternative methods of teaching, to make meaningful comparisons.

There are even fewer programmes for teaching simple number skills to retarded children and these are reviewed by Malpass (1968).

Amongst the arithmetic studies, the investigations by Cartwright (1962) and Price (1962) are of some importance. In the first study the effects of the differential arrangement of fraction programmes on learning, recall and transfer on the two groups of mentally retarded adolescents were studied. Cartwright divided the forty mentally retarded adolescents into two groups. Three main findings emerged from this study. First, the group receiving programmed instruction in systematic sequences retained more of the programmes. Secondly, the "unsystematic" programmed instruction group showed greater transfer ability and thirdly, no differences in mean learning scores were apparent between the two groups. Price (1962) in a well-designed experiment, compared the results of teaching thirty-six mentally retarded children on (a) a linear programme (b) a branching programme and (c) conventional teaching. The groups were matched for chronological and mental ages, IQ and arithmetic attainment. Two of the main findings are of interest, namely, that both the groups taught by programmed instruction proved significantly superior to the one taught by the conventional method, and that the instructional time saved was considerable (P.I.—86 lessons compared with 130 in the conventionally taught group).

Stolurow and Walker (1962) mentioned that linear programmes were more useful for dull and younger subjects. Physical activity, such as writing and response (overt responses), tended to damp down extraneous stimulation, enabling such subjects to attend to the task more easily.

In an adventurous experiment Birnbrauer (1962) "successfully" introduced programmed instruction for an entire class using available branching programmes on reading, writing, arithmetic and some "practical subjects and skills." His experiment suggests that the fully programmed classroom for mentally retarded subjects is a possibility.

PROGRAMMED INSTRUCTION FOR CEREBRALLY PALSIED CHILDREN

Very few studies have yet been reported of the use of programmed instruction with cerebral palsied children. Assessing the general effectiveness of programmed instruction is difficult because many studies are characterized by poor experimental design, unsuitable programmes, inefficient control of variables and inadequate evaluative criteria. Hartley (1966) examined eighty-four studies comparing programmed instruction with conventional teaching and on the minimal criteria for information concerning length of programme, number of students participating, amount taught, time taken and amount learned, only six studies were satisfactory. However, the evidence suggests (Leith, 1964) that though programmed instruction is not superior to conventional teaching in all respects, children taught by the former method learn in less time, retain more information and over a longer period of time (Kay et al. 1963), and have a better understanding of the subject than those instructed by conventional methods (Popham, 1964). Unfortunately, the enthusiasts of programmed instruction tend to ignore the immense practical and administrative difficulties of introducing this form of teaching into the classroom.

Hypotheses

Children suffering from cerebral palsy are commonly handicapped in learning arithmetical skills. A part of the handicap is the direct result of lesions within the central nervous system. The practical limitation of his performance cannot be predicted by a neurological examination, however, as there are secondary effects of a given defect which can be traced to a variety of social and personal reactions: some children compensate remarkably for defects which appear to be insuperable to others.

One type of variable to which all children are exposed, irrespective of physical handicap, is recognised in different forms of teaching practice. To the extent to which physical handicap limits the compensatory reactions available to a child, there is greater importance in the specific form of instruction used in the classroom.

Present teaching practice is exploring the use of programmed instruction in a number of different subjects. Using programmed instruction for cerebral palsy children has certain obvious advantages when continuity of attendance during the term is interrupted by illness or accident, or in classes with excessively wide ranges of intellectual capacity. These are situations which are particular problems in schools for physically handicapped children.

Before programmed instruction is generally recommended to aid the solution of such difficulties in schools for physically handicapped children its application in regard to the classroom needs to be compared with the usual methods of teaching. Practical difficulties due to motor handicaps, perceptual disorders etc. might render the use of programmed instruction unsuitable. The primary purpose of this study was therefore an evaluation of programmed instruction in the classroom for cerebral palsied children.

The hypotheses on which the evaluation of this study was to be based were the following:

1. Overall
   a. Cerebral palsied children can be taught by programmed instruction.
   b. Programmed instruction will prove more effective than teaching by conventional methods, especially in relation to
      (1) time taken to learn.
      (2) level of computational achievement.
      (3) degree of pupil satisfaction.
2. Individual
   Children suffering from multiple handicaps will benefit more obviously from programmed instruction than children with single disabilities, because the pace of learning is directly monitored by the child and not by arbitrary standards imposed by the classroom.

PART TWO

CHAPTER II

EXPERIMENT

SAMPLE AND PROCEDURE

THE AIMS OF THIS study as set out in the previous chapter require a comparison between two teaching methods applied to cerebral palsied children. Normal controls were not needed as this is not an attempt to compare learning progress in normal and cerebral palsied children. Controls were provided by the use of matched samples taught by programmed instruction and the usual methods of instruction. All children in both groups suffered from a common handicap, namely cerebral palsy, and the matching procedure is set out below.

The effects of variables such as the quality of instruction given by different teachers cannot be eliminated from the control group; even the same teacher is uneven in his quality of work from time to time. This is an inherent problem in classroom teaching, and its effects should therefore be retained when evaluating the result from different methods.

The nature of the school, its ethos, attitudes and composition may also have general effects on learning which affect children in specific ways, so that each of the schools selected should provide members of both experimental and control groups for comparisons.

Finally there are individual differences of children to be considered in any such assessment: these may be quite separate from intellectual attributes and result from individual likes or dislikes for a given teacher, failure to concentrate during group instruction, or perhaps from lack of persistence when left to work on their own.

In this section is presented a description of the sample and procedure used to acquire and analyse the data under the following headings:

1. Selection of sample.
2. Measurement of intellectual and emotional factors.
3. The arithmetic programme and the teaching machine.
4. Measures of arithmetic attainment.

Selection of Sample

Three schools for cerebral palsied children were selected for the experiment; their composition, location and teaching facilities were closely similar:

School A is a bilateral mixed secondary school with children of an age range of eleven to twenty-one years and an IQ range of 80 to 130.

School B is a mixed secondary modern type school with children of an age range of six to sixteen years and an IQ range of 70 to 110.

School C is a mixed secondary modern type school with an age range of seven to sixteen years and an IQ range of 65 to 110.

These children were drawn from various parts of England and Wales and have tended to include a more handicapped group than those attending local authority schools.

As mentioned earlier all the children had been diagnosed as cerebral palsied prior to admission, but though there was disagreement about the classification of handicap, some measure of consistency was secured because the same paediatrician had examined all the children in each school.

Originally two secondary modern schools (B and C) with fifty and sixty-six pupils respectively were chosen. These mixed schools were almost similar in size, ability level and age range and broadly offered the same educational opportunities. Three similar type schools were selected to ensure a sufficiency of subjects for the study in the face of such eventualities as unexpected withdrawals of pupils or crises involving prolonged surgical treatment, transfers or deaths of children. This safeguard was justified because at one stage an administrative change in one of the schools caused a serious interruption in the experiment, resulting in the loss of data on some children. A few weeks later it was decided to increase the range and numbers in the sample by including a further sixteen children from the bilateral school for cerebral palsied pupils (School A) who were slightly older children.

TABLE III
COMPARISON OF THE DISTRIBUTION OF CHILDREN IN THE EXPERIMENTAL, CONTROL AND THE COMBINED GROUPS

School	Class	Experimental (No.)	Class	Control (No.)	Total
A	1	4	1	4	15
	2	3	2	4	
B	3	4	3	3	13
	4	3	4	3	
C	5	4	5	4	14
	6	3	6	3	
Total		21		21	42

Environmental Features

Schools B and C are situated in picturesque though rather inaccessible parts of the country, and are adapted mansions set in ample grounds. With the exception of eight day pupils in one school all are residential. School A is in a modern building specially designed to overcome the barriers cerebral palsied children meet in the classroom, at recreation and in their living accommodation. All the classes of all three schools are small and well below the statutory limit recommended by the Department of Education and Science. The generous provision of houseparent and therapy staff ensures a high level of therapeutic and physical care of the pupils. There is a flexible approach to the curriculum and a judicious use of teaching and mechanical aids. Class teachers usually instruct pupils in all subjects, except in the case of specialists for advanced students.

Prior to the experiment several visits were paid to the school to observe the conditions in which the children worked. Many discussions were held with the Head and the staff to explore the expectations or anxieties teachers felt towards programmed instruction. On being invited to give their reactions, two teachers expressed some misgivings about the value of this "new method."

"I doubt if spastic children can learn well without the personal touch. . . . I personally feel we need to be sure that teaching machines do not exclude the teacher."

It was pointed out that there was no risk of spoiling contact between teacher and pupil from a daily period of forty-five minutes programmed instruction. During this period the child would indeed be working alone, but taken out of daily teaching time of over five hours, this was probably unimportant as a threat to personal relationships.

Subjects

Children were chosen who seemed likely to benefit from the programme and who had received a minimum amount of formal teaching in the subject. They were also required to have reached the chronological age of nine years, but could include others up to sixteen years. Since the only available programme at the time was originally written for non-cerebral palsied children of average and slightly above average intelligence and age eight plus, it was felt that these age limits were appropriate for this group of cerebral palsied children.

Two entire classes in each school were included to avoid disrupting the daily programme and to ensure that none of the children felt "left out" of the experiment. This arrangement also ensured that the teacher had overall and continuous control of the administration and supervision of his own class. Originally forty-five children took part, including both sexes as well as a range of age and handicaps. Each class was divided almost equally into an experimental and control group and efforts were made to match the two groups according to handicap, age and sex as far as possible. One of the teachers instructed the control group and the other teacher supervised the programmed instruction group. Only forty-two were able to complete the experiment because three children were withdrawn from school for prolonged hospital treatment.

MEASUREMENT OF INTELLECTUAL AND EMOTIONAL FACTORS

As was demonstrated in the review of literature there are many factors which help or hinder normal children in learning and that even wider considerations apply to the cerebral palsied child. There are studies which show that sensory, intellectual and emotional factors may all play a part in such learning. Information about the children's sensory disorders (hearing and vision) was available in each case. All the children taking part in the study were given the following tests. Except for the WISC (administered by the author), the tests were administered by the teachers, who were especially trained for this purpose. The author was responsible for giving and evaluating all test results subsequently. Tests were given at intervals over a period of weeks preceding the beginning of the experiment.

Intellectual Factors

1. Compound Series Test: Morrisby (1955).
2. Wechsler Intelligence Scale for Children: Wechsler (1949).
3. Goodenough Draw-a-Man Test: Goodenough (1926).
4. Bender Visual Motor Gestalt Test: Bender (1938).

Emotional Factors

5. Draw-a-Person Test: Machover (1949).
6. Bender Visual Motor Gestalt Test: Bender (1938).
7. Bristol Social Adjustment Guide. The child in residential care: Scott (1963).

THE ARITHMETIC PROGRAMME AND THE TEACHING MACHINE

Choosing the Machine and Programme

After attending the demonstrations of various teaching machines at exhibitions, firms and schools, a machine using a linear programme was selected. Several considerations influenced this choice. Safety and cost were of special importance.

Branching machines were prohibitively expensive and often needed electrical and mechanical attachments, which could prove hazardous for children using crutches or wheelchairs. These machines were likely to break down and hold up the progress of this type of experiment at a crucial stage. Most of the branching programmes were expensive and the few available ones were unsuitable for the children. It also seemed that the branching programme required much reading before a choice between answers was made.

A final consideration was the marked difference in positive reinforcement provided. In the branching programme the child is likely to make many mistakes, but the linear programme is designed to ensure that the child almost always obtains the correct answer. This is an important consideration with children who have so often had to face failure. By enjoying almost constant success, as they worked through the linear programme, it was expected the children would regain confidence in learning arithmetic.

The ESA tutor machine and the "Primary Arithmetic" programme written by R. D. Bews was chosen. The machine and the programme were obtained from the Educational Supply Association Ltd. Because of the gross physical handicap and lack of fine finger dexterity of some athetoids and spastics, some structural modification to the machine was required. It was discovered that some athetoids were "overshooting" the hold and not exposing each frame in the correct order. A simple and effective handle was designed to compensate for the exaggerated or weak hand movements. The machines were also mounted on adjustable stands enabling the children to work with comfort and confidence.

The Programme

The "Primary Arithmetic" programme consists of ten sets with ninety or more instructional frames in each set. At the end of each set a test measures the degree to which the child has assimilated what he has been taught. The programme is presented on stiff cardboard sheets. Each sheet contains six

frames. Normal children need between twenty-four and twenty-eight hours to complete the programme.
Each set is briefly described in Table IV.

TABLE IV
"PRIMARY ARITHMETIC" PROGRAMME

			Number of Frames	Approx Time in Hours	Test Card Items
Set	2	Subtraction by Equal Addition	90	2-3	8
Set	3	Subtraction by Decomposition	90	2-3	8
Set	4	"Nought difficulty	90	3	8
Set		(Sub by Equal Addition)			
Set	5	"Nought" difficulty	90	3	8
Set		(Sub by Decomposition)			
Set	6	Multiplication	90	2-3	20
Set	7	Division	90	3	24
Set	8	Equivalence	90	2-3	24
Set	9	Fractions	138	3	17
Set	10	Decimal Fractions	90	2	16

Administration of the Teaching Programme

Teachers

In each school, two teachers shared their time between the experimental and control groups, spending a session with one group and the next with the other group. In this way it was hoped the influence of different teaching styles and of the personality of the teachers would be minimised.

Times

Both groups had five thirty minute sessions per week. The experimental group children could break off whenever they pleased and the time they had spent on the machine was recorded.

Syllabus

After several trial runs to familiarize themselves with the operation of the machines, the experimental group worked through sets two to ten of the programmes at their own pace. The control group were taught the same subjects and processes in whatever way the teacher thought appropriate. The teacher administered the attainment test card at the end of each set when it was felt the control group was ready.

Recording Results

Three kinds of results were collected:
1. Number of frames correctly and incorrectly answered.
2. Scores on the attainment tests given at the end of each set.
3. The reactions of the children and the teachers as the programme was worked through.

Before they began the next card of six frames, the children inserted a fresh answer sheet. As soon as that card was completed the answer sheet was removed and the child would immediately know how many of his answers had been correct. These results could of course only be collected from the experimental group.

The attainment tests were marked immediately, so the children knew their score before the end of the lesson. These tests were taken by both the experimental and the control groups.

The teachers kept records of each child's progress. They noted when the children had difficulty with the programme or with operating the machine. They also recorded spontaneous comments, reactions and difficulties; on completion of the programme the teachers were also encouraged to report their own comments and feelings about the merits and faults of the machines and programme.

MEASURES OF ARITHMETIC ATTAINMENT

In order to observe differences in postlearning attainment between the experimental and control groups two arithmetic tests were chosen:

1. Southend Attainment Tests in Mechanical Arithmetic (Southend, 1939).
2. Schonell Diagnostic Arithmetic Tests (Schonell and Schonell, 1960).

Southend Attainment Test in Mechanical Arithmetic (Southend)

This test is designed to survey arithmetic attainment between the ages of six and fourteen years. A separate subtest covers each age group. Only the four subtests (see Table V) covering the ages of six to ten years were used because these sampled the entire range of attainment achieved by the children.

Experiment

TABLE V
SOUTHEND SUBTESTS

Age Group	No. of Items	Content
6-7 years	10	Addition + subtraction
7-8 years	10	Subtraction and multiplication
8-9 years	10	Division, multiplication and money addition and subtraction
9-10 years	5	Division and money addition, subtraction, division and multiplication

The test was given to the experimental and control groups immediately after the children had completed the programme. An overall score is obtained which is expressed as an arithmetic age. The test consists of only a few items but it does provide a quick review of arithmetic progress.

Schonell Diagnostic Arithmetic Tests (Schonell)

This test battery is made up of twelve tests. As each test contains a large number of items the results are more reliable and permit finer discriminations than the Southend test. Except for tests 5 and 12, each test deals with only one of the four processes. It is therefore possible to identify specific difficulties in grasping any of the four rules. The twelve tests are briefly described in Table VI.

These tests were administered by the teachers over several sessions after the children had completed the programme. Schonell provides arithmetic age norms for timed tests, but all the tests were given untimed as the level of understanding achieved was more important than speed.

TABLE VI
SCHONELL DIAGNOSTIC TEST BATTERY

Test	No. of Items	Content
1	100	Addition
2	100	Subtraction
3	100	Multiplication
4	90	Division
5	100	Four rules
6	58	Graded addition
7	56	Graded subtraction
8A	42	Graded multiplication (easy steps)
8B	11	Graded multiplication (harder steps)
9	44	Graded division
10	36	Graded long division (easy steps)
11	24	Graded long division (harder steps)
12	40	Graded mental arithmetic

CHAPTER III

RESULTS

STATISTICAL ANALYSIS OF THE RESULTS

Computer Analysis

FORTY-FIVE CHILDREN took part in the experiment. The results for three of them were incomplete on account of periods of illness or because they left the school. For the remaining forty-two children results were available on fifty-seven variables.* They were analysed on the I.B.M. computer which provided means, standard deviations and product moment correlations for the entire sample. The distribution of a number of these variables deviated from the normal curve, but not sufficiently to invalidate the use of robust techniques of analysis.

Means, standard deviations and "t" tests of significance for the difference between the experimental and the control groups and for the differences between the three schools on each variable were also obtained.

Manual Computation

Where appropriate, partial correlations were calculated from nomographs provided by Educational Testing Service (Research Bulletin, 1960). X^2 and the Mann Whitney "U" tests were applied where there was data for fewer than the forty-two subjects or where comparisons were required between the research group and the rest of the schools sampled.

COMPARISON BETWEEN EXPERIMENTAL AND CONTROL GROUPS

The presentation of the results in this section will be made in

---
* Note: For a complete list of variables see Appendix.

five subsections. The experimental and control groups will be compared on the following:
1. Age, sex and handicap (descriptive data).
2. Intelligence tests, Bender Gestalt and Stott (predictor variables).
3. Arithmetic attainment (criterion variables).
4. Programme instruction, set tests.
5. Learning time.

Descriptive Data

Age

The results are shown in Table VII. From this it may be seen that there were no significant differences in age between the two groups.

TABLE VII

MEAN CHRONOLOGICAL AGE AND SD OF EXPERIMENTAL AND CONTROL GROUPS

	Experimental (N=21)	Control (N=21)
Mean CA (yr)	12.7	12.6
SD	2.2	2.04

Sex

There were almost twice as many boys as girls in both groups. No significant differences in the sex distribution of the two groups were revealed, χ^2 being 0.17.

TABLE VIII

SEX INCIDENCE IN EXPERIMENT AND CONTROL GROUPS

	Experimental (N=21)	Control (N=21)	Total
Boys	15	14	29
Girls	6	7	13
	21	21	42

$$X^2 = 0.17$$
$$\text{"t"} = 0.1$$

50 Arithmetical Disabilities in Cerebral Palsied Children

Handicap

Table IX gives the results on the distribution of handicap in the experimental and control groups. There were no significant differences in the distribution of athetoids, diplegics and spastics (other than diplegics) between the two groups ($\chi^2 = 3.33$).

TABLE IX
DISTRIBUTION OF HANDICAP IN EXPERIMENTAL
AND CONTROL GROUPS

	Experimental (N=21)	Control (N=21)	Total
Athetoid	7	9	16
Diplegic	6	8	14
Spastic (Other Than Diplegic)	9	3	12
	22	20	42

$X^2 = 3.33$ (NS)

Predictor Variables (9-44)

Intelligence Tests (9-19)

Table X shows the mean scores, standard deviations and ranges of the intelligence tests for the experimental and control groups. It can be seen that both groups were carefully matched on all these variables and that there were no significant differences on any of the measures.

TABLE X
COMPARISON BETWEEN EXPERIMENTAL AND CONTROL GROUPS OF
MEAN, SD AND RANGE OF INTELLIGENCE TESTS

Variable		Experimental (N=21) Mean	SD	Range	Control (N=21) Mean	SD	Range	"t"
9	WISC Full IQ	87.81	14.13	67-125	85.43	15.92	54-120	0.5
10	WISC Verbal IQ	96.57	16.16	76-135	95.33	16.52	65-137	0.2
11	WISC Perf IQ	80.38	17.00	53-115	76.52	17.31	48-113	0.7
12	CST IQ	82.10	18.16	62-134	77.57	15.90	62-108	0.9
13	DAM IQ	54.33	11.49	31-74	54.62	14.16	30-75	0.1
14	WISC "Verbal"	28.57	9.28	15-49	28.14	8.95	12-49	0.2
15	WISC "Spatial"	13.19	6.70	1-30	11.52	5.06	1-23	0.9
16	WISC Information	9.81	3.09	5-15	8.91	2.64	4-15	1.0
17	WISC Arith	8.76	3.10	4-17	8.91	3.29	5-15	0.1
18	WISC Pic Comp	8.19	2.77	4-15	8.76	3.43	3-16	0.6
19	WISC Pic Arr	7.81	3.01	4-15	6.24	3.02	1-12	1.7

Discrepancy Scores (20-24)

The discrepancy scores of the various intelligence tests and their subtests of the experimental and control groups is presented in Table XI. It will be noted that there were no significant differences between the groups in this respect.

TABLE XI
COMPARISON BETWEEN EXPERIMENTAL AND CONTROL GROUPS OF MEAN, SD AND RANGE OF DISCREPANCY SCORES

Variable		Experimental (N=21) Mean	SD	Range	Control (N=21) Mean	SD	Range	"t"
20	WISC IQ (V-P)	16.19	16.76	1-59	18.81	14.20	1-49	0.4
21	WISC IQ (VR-K)	15.38	10.41	1-43	16.62	8.51	3-36	0.5
22	WISC (V-DAM)	42.24	20.63	5-79	40.71	18.29	11-70	0.3
23	WISC (P-DAM)	26.10	14.47	3-60	21.91	14.82	4-52	0.9
24	WISC (Full-DAM)	33.48	13.72	10-54	30.81	15.58	4-58	0.4

Bender Gestalt (Variables 25-32)

The Means, S.Ds and Ranges of the Bender Gestalt test scored according to the Bender and Koppitz system are compared in both groups (see Table XII). Except for Variable 26, none of the differences on the other measures between the experimental and control groups reached significance levels.

TABLE XII
COMPARISON BETWEEN EXPERIMENTAL AND CONTROL GROUPS OF MEAN, SD AND RANGE OF BENDER GESTALT SCORES

Variable		Experimental (N=21) Mean	SD	Range	Control (N=21) Mean	SD	Range	"t"
25	BG Full Score	37.24	9.01	9-01	34.76	8.34	7-47	-0.9
26	Koppitz Distortion	18.38	2.92	11-22	16.57	2.32	11-21	-2.2*
27	Rotation	1.71	0.46	1-2	1.71	0.46	1-2	0.0
28	Integration (Total)	16.33	2.20	9-18	15.33	2.87	9-18	-1.3
29	Integration Fig. 4	1.67	0.48	1-2	1.67	0.48	1-2	0.
30	Integration Fig. 3	1.57	0.51	1-2	1.48	0.51	1-2	-0.6
31	Integration Fig 5	1.81	0.40	1-2	1.67	0.48	1-2	-1.0
32	Perseveration	1.67	0.48	1-2	1.62	0.50	1-2	-0.3

* t Sig. at 5% level

Measured by the "t" test the differences between the mean scores of the experimental and control groups on Koppitz Distortion Score are shown to be significantly different ($p<.05$).

Bristol Social Adjustment Guide (Stott)

Table XIII shows the differences between the mean scores, S.Ds and ranges of the scores of the experimental and control groups on the various items of the Stott test. It can be seen that there were no significant differences between the scores of the two groups.

TABLE XIII

COMPARISON BETWEEN EXPERIMENTAL AND CONTROL GROUPS OF MEAN, SD AND RANGE OF BRISTOL SOCIAL ADJUSTMENT GUIDE (STOTT)

Variable		Experimental (N=21) Mean	SD	Range	Control (N=21) Mean	SD	Range	"t"
33	Total No Symptoms	12.05	8.36	4-29	10.14	5.83	3-25	0.9
34	Graded symptoms	1.67	0.80	1-3	1.62	0.67	1-3	0.2
35	U+W	2.24	2.26	0-7	1.76	1.30	0-5	0.8
36	D	1.29	1.45	0-5	1.48	1.08	0-4	0.5
37	XA	1.95	1.99	0-7	1.71	1.35	0-4	0.5
38	HA	1.81	2.48	0-9	1.14	1.93	0-7	1.0
39	K	0.52	1.03	0-5	0.38	0.92	0-4	0.5
40	XC	0.48	0.93	0-3	0.57	0.87	0-3	0.3
41	HC	0.24	0.77	0-4	0.24	0.54	0-3	0.0
42	R	1.43	1.81	0-6	0.86	1.20	0-5	1.2
43	M	1.52	1.29	0-4	1.33	1.02	0-3	0.5
44	MN	0.57	0.75	0-2	0.67	0.80	0-2	0.4

Arithmetic Attainment

Preliminary Assessment

Since the children were drawn from three separate schools it was especially important that their arithmetic attainment prior to the experiment should not only be known but also lie within a relatively narrow range.

Arrangements were made therefore at each of the schools for a preliminary test of arithmetic attainment to be administered to each child before the experiment commenced. Although this was carried out at one of the schools, teachers at the other two schools had begun their teaching programme prematurely and had not allowed for preliminary testing. Accordingly the scores of the first set results are taken in place of the initial attainment scores. This is legitimate as less than three sessions had been completed at the time of this test. It would have been preferable to have used an arithmetic attainment test even now, but as about half the pupils had completed their attainment test before the onset of the experiment, they could not have been compared directly with attainment tests given after the experiment had begun.

An obvious advantage of using the results from the first set test as a basis of comparison between the experimental and control groups is that both groups share a common learning experience, irrespective of the teaching methods adopted by individual teachers in three schools. In a group of this size, individual differences of training could play a major role in determining test results so that any measure of standardisation is helpful for purposes of comparison.

Table XIV presents a comparison between the experimental and control group of the arithmetic attainment tests. It will be seen that on both the Southend and Schonell Arithmetic Tests as well as various subtests of the Schonell, no significant differences in the mean scores of the groups emerged.

TABLE XIV

COMPARISON BETWEEN EXPERIMENTAL AND CONTROL GROUP OF MEAN, SD AND RANGE OF ARITHMETIC ATTAINMENT TESTS

Variable		Experimental (N=21) Mean SD Range			Control (N=21) Mean SD Range			"t"
45	Southend	80.19	6.73	67-96	77.86	11.07	41-98	0.8
47	Schonell–Total	520.48	153.39	238-769	481.76	206.47	64-783	0.7
48	Schonell–1-5	368.57	100.40	118-532	388.57	74.25	168-522	0.5
49	Schonell–8-11	45.33	43.35	0-143	60.48	81.69	0-344	0.8
50	Schonell–12	16.57	10.77	0-34	18.67	13.92	0-37	0.5
51	Schonell–3	75.81	20.59	39-99	64.33	32.27	0-99	1.4
52	Schonell–4	61.10	26.24	0-90	57.95	31.48	0-90	0.4
53	Schonell–5	63.86	29.94	0 00	59.33	36.15	0-99	0.4
54	Schonell–6	42.33	17.68	0-57	37.14	20.18	0-58	0.0
55	Schonell–7	31.00	18.58	0-55	30.29	19.24	0-55	0.1
56	Schonell–8a	16.05	12.75	0-41	13.19	13.86	0-41	0.7
57	Schonell–9	19.76	13.21	0-41	18.19	15.0	0-43	0.4

Programme Instruction, Set Tests

Table XV gives the results comparing the Programme Instruction set test scores for the experimental and Table XVI for the control groups in Schools A and B. The results of School C were

54 Arithmetical Disabilities in Cerebral Palsied Children

TABLE XV
COMPARISON BETWEEN EXPERIMENTAL AND CONTROL GROUP OF
SET TEST SCORES ON ARITHMETIC PROGRAMME

Experimental

	Pupil	Set 2	3	4	5	6	7	Total	Percentage
School A	1	3	1	7	3	20	13	47	69.11
	2	8	8	6	7	20	15	64	94.11
	3	6	8	8	8	17	15	62	91.17
	4	8	7	6	7	16	16	60	88.13
	5	8	8	8	8	20	16	68	100.00
	6	8	8	8	7	20	16	67	98.52
	7	8	8	7	6	20	11	60	88.23
School B	1	6	5	7	5	15	14	52	76.47
	2	8	8	8	8	20	15	67	98.52
	3	3	5	3	6	20	10	47	69.11
	4	6	8	7	1	20	11	53	77.94
	5	8	7	8	8	20	16	50	73.52
	6	8	7	6	7	18	12	58	85.29
	7	8	8	6	7	20	15	64	94.11
Mean		6.86	6.86	6.79	6.29	19.00	13.93	58.50	85.71
SD		1.8	2.0	1.4	2.1	1.8	2.1	7.4	11.0

TABLE XVI

Control

	Pupil	Set 2	3	4	5	6	7	Total	Percentage
School A	1	8	8	8	8	20	15	67	98.52
	2	1	0	2	2	8	5	18	26.47
	3	8	5	8	7	17	10	55	80.88
	4	7	8	8	8	20	16	67	98.52
	5	8	8	8	8	17	16	65	95.58
	6	8	7	8	8	18	16	65	95.58
	7	8	8	8	8	20	13	65	95.58
School B	1	7	7	7	6	20	15	62	91.17
	2	8	8	3	6	20	14	59	86.76
	3	7	8	8	8	20	14	65	95.58
	4	8	8	8	8	20	12	64	94.11
	5	3	8	7	4	20	13	55	80.88
	6	6	7	2	6	18	5	44	64.70
	7	4	7	2	2	20	11	46	67.64
Mean		6.50	6.93	6.21	6.43	18.93	12.50	56.93	83.14
SD		2.2	2.5	2.6	2.1	3.2	3.7	13.5	19.8

incomplete and therefore not included in the analysis. Also as sets 8, 9 and 10 were not attempted by all the children, these test results were not analysed. Combining set test scores for two and three we find that there were no significant differences between the experimental and control group in either schools A or B.

On the Mann Whitney U Test, School A p = 0.446 (2 tailed test) and on the Mann Whitney U Test, School B p = 0.866 (2 tailed test) comparing scores on set test 7 in School A, the experimental groups were almost significantly superior, p = 0.116 (1 tailed test). In School B, the scores of the experimental group were significantly superior to those of the control group, p = 0.026 (1 tailed test).

As demonstrated by the "t" tests neither the average set test scores nor the overall results differed significantly between the two groups. On the other hand an inspection of the variances of the two samples shows that the variance in the control group is consistently higher than that for the experimental group. In other words the range of individual differences in attainment obtained by conventional teaching methods is consistently greater than that obtained by programmed instruction. Furthermore this tendency to emphasise individual differences is progressive

TABLE XVII

COMPARISON BETWEEN EXPERIMENTAL AND CONTROL GROUPS OF MEAN AND SD OF SET TESTS SCORES ON ARITHMETIC PROGRAMME

		Sets 2	3	4	5	6	7	Total	Percentage
Experimental Group (N=14)	Mean	6.86	6.86	6.79	6.29	19.00	13.93	58.50	85.71
	SD	1.8	2.0	1.4	2.1	1.8	2.1	7.4	11.00
Control Group (N=14)	Mean	6.50	6.93	6.21	6.43	18.93	12.50	56.93	83.14
	SD	2.2	2.5	2.6	2.1	3.2	3.7	13.3	19.8
	"t"	0.46	0.09	0.72	0.18	0.58	1.26	0.38	0.45

throughout the series. It could therefore be argued that the conventional teaching method is a less reliable way of teaching than programmed instruction at least for the purposes of this experiment.

Learning Time

Table XVIII compares the times taken to complete the programme by each child in Schools A and B in the experimental group. It is interesting to note the wider range in time taken

TABLE XVIII

COMPARISON OF LEARNING TIME (IN MINUTES) TO COMPLETE ARITHMETIC PROGRAMME BY SCHOOLS A AND B IN EXPERIMENTAL GROUP

		School A Sets 2-7	School B Sets 2-7
Subjects	1	710	646
	2	420	702
	3	1,280	703
	4	640	591
	5	530	600
	6	270	623
	7	480	608
Total		4,330	4,473
Average		618.6	630
Range		270-1,280	591-703

to complete the programme in School A as compared with School B. This adds emphasis to the plea that cerebral palsied children should be allowed to set their own learning pace rather than be taught at administratively convenient rates.

There were no marked differences in instructional time for the schools in the control group. In School A the children taught by conventional methods attended twenty-four periods of arithmetic instruction, each lesson taking thirty minutes (total 720 min), whereas children in School B received twenty-three lessons (total 690 min).

PROGRAMMED INSTRUCTION GROUP ALONE

Errors

Table XIX shows the correct responses expressed in totals and percentages by Schools A and B in the Experimental Group. It can be seen that only three children out of fourteen satisfied the minimum criteria of 95 percent correct responses to the linear programmes.

Teachers' Comments About Programmed Instruction and Teaching Machines

In this section a sample of teachers' comments and their reaction to programmed instruction elicited during informal discussion and by written responses are offered. Care has been

TABLE XIX
COMPARISON OF CORRECT ANSWERS TO PROGRAMMED INSTRUCTION BETWEEN BOTH SCHOOLS IN EXPERIMENTAL GROUP

		Sets	2	3	4	5	6	7	Total	Percentage
	Pupil	Maximum Score	90	90	90	90	90	90	540	100
School A	1		81	70	68	78	87	60	444	82.03
	2		90	87	33	88	85	84	467	86.48
	3		85	78	80	88	86	87	504	93.30
	4		90	82	88	88	87	81	516	95.50
	5		89	89	88	90	89	89	534	98.80
	6		82	86	90	89	88	89	524	97.03
	7		81	83	81	77	90	85	497	92.03
School B	1		61	75	74	79	79	65	433	80.18
Subjects	2		84	84	86	89	83	81	507	93.20
	3		63	75	66	65	68	72	409	75.74
	4		79	86	84	84	89	89	511	94.62
	5		81	80	85	87	87	84	504	93.30
	6		86	74	78	83	81	75	477	88.30
	7		86	81	85	88	88	82	510	94.40

exercised to avoid identification of the respondents and in general the more critical views are reflected. Not all the comments are selected from one school, though a disproportionately greater number of statements were received from School B.

Where possible these responses are categorised, but it was felt that the spontaneity of their feelings should not be affected by too rigid a classification. It will become evident that some repetition and contradiction in their statements occurs which bears testimony to the unrehearsed style of expression. A standardised attitude questionnaire might have inhibited freer comments on programmed instruction. It also appears that on balance the children reacted more favourably to programmed instruction than did the teachers and this was honestly reported by teachers.

Pupil Satisfaction

A serious objection to programmed instruction raised by one teacher was that the reward from "immediate feedback" informing children of the correctness of their answers was inadequate. Furthermore a number of social and interpersonal contacts "which are taken for granted in classroom teaching" were curtailed by this form of instruction.

Some students resented spending the time loading the machine, and were also irritated by the mechanical difficulties of operating the instrument. With a few severe athetoids this triggered off excessive dribbling on the machine, which increased handling problems "when loading the answer papers because the metal was wet."

Two teachers independently commented on the general interest and satisfaction derived by pupils working on the machine. One remarked "in each and every case the pupils stated they enjoyed working with the machines" but added a footnote "two of these pupils told a different story from day to day." Another teacher wrote "all enjoyed it except one who thought it was a waste of time."

Amongst other comments offered by teachers was the satisfaction pupils experienced in "working at their own pace," the appeal and novelty of "a machine" for young boys and the feeling of "independence" enjoyed by a few severely handicapped children who regularly required a great many concessions and were so thrilled at being able to work on their own that improvement in their "behaviour and attitude" was noted.

Teacher Satisfaction

An analysis of the wide range of responses to programmed instruction by teachers indicated that in general they commented on three aspects:

1. Machine.
2. Programme.
3. Nonacademic, personality and social matters.

Machine. Criticism of the mechanical efficiency of the machine was made by a few teachers. Such problems as "jamming of the paper . . . card slipping . . . small perspex window" and time taken up in setting up the equipment were raised. One teacher who strongly resented the extra work involved in this method of instruction said "can't really fill up with cards and leave the child to work by itself for a long period—if you are going to check whether he can do them or not."

Programme. Some teachers drew attention to weaknesses in the arithmetic programme. Generally they referred to the content and the technique of presentation. In the former category the inappropriateness of the level of difficulty was mentioned, though it was interesting to note how the teacher, during a pilot run, "misunderstood the instructions" and tried out the programme on a small group of "bright 16 and 17 year old pupils" and suffered throughout the entire period, angrily declaring that the "students were bored and puzzled at being taken through this very elementary course in arithmetic."

Among some suggestions offered were the need for more drill, smaller number of steps in some cases and better method of presentation. Sample reactions are offered:

". . . if they were in sets that taught certain processes and gave plenty of practice then I think they would be of value to a child who, finding number work difficult, dislikes it but is encouraged to do it because of the appeal of the machine."

". . . proceeds too quickly with new processes. The child is not given a chance to practice what he has just learnt."

". . . if the child puts the cards in, he can see the answers before he starts."

"I feel that the tests with the machines have not proved the effectiveness or otherwise of these machines. The children using them had, in the majority of cases (this is not to suggest that the children were competent) previous knowledge of the subject matter on the subject. It is very doubtful whether one could say whether the children were absorbing the teaching instructions then answering the questions or just answering the questions without any regard to the machine (i.e. in spite of the machine).

In some cases the pupils suggested that the instructions on the cards had confused their minds and they pointed out differences (or rather 'imagined' differences) between the teachers' ideas and the machine's ideas. I would like to see the machines used again, but this time with material absolutely new to the pupil—such as 'set theory' in modern mathematics."

It was clear that one teacher felt that the children would have worked better on the programme, had they felt free to call on the teacher for help with frames that gave them trouble. Some children were reported to "ignore the information or advice on the cards and just answer the question framed in the last sentence, hence a correct answer does not necessarily mean that the student has been 'taught' by the machine." It is not clear whether the teacher felt that completion of sections of the programme was a poor guide for checking whether the children had understood it.

Some valuable suggestions were offered for extending its use for children with other disabilities, such as the deaf.

> I do feel, however, that the machines, if modified mechanically, have a future, particularly with regard to
> (a) the hard-of-hearing who may, and often do, miss much of the teacher's verbal instruction
> (b) the very heavily handicapped who are, of necessity, left alone for considerable periods during lesson time. With these pupils the machine would have to be electrically operated and the answers, of a Yes/No variety, would be indicated by the pressing of the appropriate button. (Unless of course, an amanuensis was employed, but this would, I think, defeat the object of using a machine).

Nonacademic Factors. Some teachers were unhappy at handing over control of teaching to a machine. They were disenchanted about being relieved from unnecessary chores. . . .

"I thought that I would have more time to spend with the children, not on the machines. Actually I have not had as much time and I felt it to be a continual rush. I see no advantage in using the machines."

The "reduction of personal contact" with children was regretted and safeguards for ensuring "warm human relationships" with the children were insisted upon.

PREDICTORS OF ARITHMETIC SUCCESS

Reduced Correlation Matrix

Originally data on the intercorrelations of fifty-seven variables were available. Those variables which gave significant correlations only were selected to be included in the reduced matrix shown in Table XXII following. As can be observed from the reduced correlation matrix, certain variables tended to have similar size correlations with the subtests of the Southend Arithmetic Test. It seemed more appropriate to use the Southend Arithmetic Test because it gave an overall level of arithmetic attainment. The Schonell on the other hand, gives results on easy, difficult and particular processes. Whilst it is interesting and important to study differential test predictors which correlate with success on one level and not another, the numbers were too small and the sample too highly selective to be of value. Despite its brevity, the Southend Arithmetic Test was highly reliable in this group. Its odd-even reliability was .93 corrected by the Spearman Brown Prophecy formula.

Table XX gives correlations of significance of Southend Arithmetic Test and other variables.

TABLE XX

SIGNIFICANT CORRELATIONS OF SOUTHEND WITH OTHER VARIABLES

Variable		Southend 45
12	CST	.49*
13	Koppitz Integ Fig. 5	.43*
32	Perseveration	.43*
17	WISC Arith	.42*
5	Diplegic	−.38†
2	Age	.34†
30	Koppitz Integ Fig. 3	.32†
4	Athetoid	.31†
25	Bender Score	.27

* Denotes significance at 1% level.
† Denotes significance at 5% level.

With regard to the other intercorrelations of the matrix it was noted that the first four descriptive variables did not reveal any specific associations with the test variables. This applies equally to age, sex, athetoid and diplegic characteristics of the children. The test variables, 6 to 55, showed their highest intercorrelations within a given test, i.e. the various items taken from the WISC, or those from the Schonell group, while intercorrelations between items taken from different tests were consistently lower.

TABLE XXI

REDUCED CORRELATION MATRIX

	3	4	5	9	10	11	12	14	15	17	25	28	30	31	32	45	47	48	49	50	54	55	
2. Age	04	04	—24	—28	—17	—34	—02	—22	—46	—18	—29	—11	03	06	—06	34	45	40	45	47	43	30	2
3. Sex		21	—07	—16	—24	—04	24	—23	08	—21	07	09	27	31	11	17	25	30	10	—03	30	16	3
4. Athetoid			—61	18	—02	33	59	—11	23	28	—11	07	31	10	24	31	31	26	38	22	34	29	4
5. Diplegic				11	10	—28	—33	20	—22	—03	—08	—23	—28	—28	—20	—38	—36	—34	—31	—25	—45	—36	5
6. WISC Full IQ					85	83	34	76	70	69	29	43	28	26	42	18	10	05	15	36	06	31	9
10. WISC Verbal IQ						41	13	94	29	73	10	22	05	07	25	15	14	10	15	37	02	34	10
11. WISC Perf IQ							45	32	92	45	40	52	43	37	46	15	03	—02	10	23	08	28	11
12. CST IQ								—06	34	41	39	45	58	33	42	49	41	34	47	31	26	31	12
14. WISC "Verbal"									21	56	—02	13	01	—01	14	—03	—03	—05	—04	23	—09	08	14
15. WISC "Spatial"										40	44	49	44	36	39	07	02	02	—01	09	11	21	15
17. WISC Arith											27	26	24	14	28	42	42	34	47	57	25	51	17
25. BG Score												77	47	58	38	27	10	09	09	08	01	27	25
28. Koppitz Intergration Total													63	73	56	30	22	21	22	15	16	33	28
30. Integ Fig. 3														52	38	32	35	37	24	23	39	25	30
31. Integ Fig. 5															57	43	39	39	30	32	44	37	31
32. Persev																43	21	17	26	17	31	35	32
45. Southend																	67	60	66	70	60	69	45
47. Schonell Total																		97	85	74	84	80	47
48. Schonell 1-5																			70	62	87	74	48
49. Schonell 8-11																				77	57	73	49
50. Schonell 12																					60	74	50
54. Schonell 6 (Graded add)																						73	54
55. Schonell 7 (Graded sub)																							

Expectancy Table

In order to determine how far along the CST (12) could pick out good and poor performances in arithmetic, the CST (12) scores and Southend (45), arithmetic ages were divided as near as possible into four quartiles, as shown in Table XXII below:

It will be observed that thirty-one children (73.8%) were correctly placed, e.g. in the top half for both CST and Southend

TABLE XXII

Southend (45)		CST (12) 62 and under	66-78	81-89	93 +	
Arithmetic ages	8.5 yrs	0	3	1	5	9
	8-0 to 8-4	0	3	7	2	12
	7-5 to 7-9	7	3	1	1	12
	7-4 and under	5	1	2	1	9
						42

and bottom half for both CST and Southend. Six (14.3%) however, were placed in the bottom 50 percent for CST, though their attainment on Southend was in the top half. In other words, they did well on the criteria test despite their poor results on the CST; while five (11.9%) children were placed in the opposite direction, e.g. their achievement on the Southend was low despite a high CST performance.

SET TEST RESULTS OF THE EXPERIMENTAL AND CONTROL GROUPS

The scores of the children on Set tests 2 to 7 are assembled in Table XXIII. Those for the last three Sets were only available for some of the children who progressed beyond Set 7 and are not included in the table.

It will be noted that the means of the two groups are closely similar, since this was a condition of the work, i.e. that a 95 percent scores should be obtained when a given set was completed. The major difference between the two groups is in the variances, which are consistently larger for the control than for the experimental group.

TABLE XXIII
MEANS AND VARIANCES FOR CORRECT RESPONSES AND ERRORS MADE BY THE TWO GROUPS IN THE SET TESTS

Correct:	Set	Experimental M	Variance	Control M	Variance	F Ratio	
	2	6.86	3.36	6.50	5.04	1.5	
	3	6.86	3.98	6.93	4.69	1.2	
	4	6.79	1.87	6.21	6.95	3.7	5%
	5	6.29	4.22	6.43	4.26	1.0	
	6	19.00	3.07	18.42	10.41	3.4	5%
	7	13.93	4.53	12.50	13.50	2.9	5%
		58.50	55.50	56.92	180.99	3.3	5%
%Correct:		85.71		83.14			

CHAPTER IV

DISCUSSION

Suitability of Programmed Instruction for Cerebral Palsied Children

THE WIDESPREAD USE of teaching machines in schools for normal children raises the question whether similar procedures can also be applied to physically handicapped children. Some theoretical and practical issues which surround this approach for the cerebrally palsied have been outlined in the introductory sections to show that past finding would predict an uncertain outcome of teaching by machine in this group of children. A practical study was therefore undertaken in which cerebrally palsied children taught by conventional methods were compared with an experimental group taught by machine. Since the results from both methods were equally satisfactory, the initial questions were answered in the affirmative: *Cerebrally palsied children are able to manipulate a teaching machine, and the progress of their learning is sufficiently rapid for practical classroom use.*

Time Taken to Learn and Level of Computational Achievement

Having established these major points, a number of contributing factors need to be considered. Taking these in the order of initial presentation, there is first the matter of the time spent by the pupil in attaining his achievement. It was apparent that the overall time spent with the teaching machine was less than that offered by standard instruction (a mean difference of 152.4 min.). In terms of gross results, therefore, the teaching machine represents a saving of time as compared with standard methods, and this is in line with the predictions. There could be many reasons for this. Perhaps the simplest is that the teaching machine offers a form of individual instruction which can rarely be attained in the classroom—even with the small groups which teachers had in the present study. Difficulties encountered by different children following a course of instruction vary from one individual to another, and the teaching machine permits a more individual rate of progress than can be attained in any classroom setting. While the overall level of attainment at the end of term was similar in the two groups, the individual progress of the children differed in the course of time. A clear sign of individual variations was reflected in the performance of the children on the tests administered at the end of each stage of the programme. Inspection of Table XXIII shows that there was a steadily maintained difference between the outcome of programmed instruction and conventional teaching methods test by test. The variances of the means and the scatter therefore of individual attainment levels in the control group was nearly three times greater than for children in the experimental group. Children taught by programmed instruction showed far greater consistency in their level of attainment and this is precisely what the experiment had expected to show. This clearly demonstrated that more homogeneous results can be expected from programmed instruction than from conventional teaching methods, and in view of the uncertainties that bedevil teaching in special schools, the reliability of instruction by programmed instruction is no mean advantage.

Pupil Satisfaction: Teacher Response

A further point raised at the beginning of the study, refers to "the degree of pupil satisfaction." An attempt was made to assess this from the teachers' reports. It was considered relevant to make such an attempt, as motives to learn could well influence the progress made by the children as much as the methods used for their instruction. Some of the incentives for using the machine were noted in the introduction: novelty value, and independence and responsibility are equally valued by the handicapped and the normal child. In addition, it was pointed out that the neutral emotional climate offered by work with a machine could be a practical advantage for cerebrally palsied children, who were so often faced by failure in everyday schoolroom tasks. Teachers commented on the former but not on the latter aspects of the machine. Some also made it clear, however, that a teaching machine was a threat to their relationship with the child, and made hostile comments to this effect.

Suitability of Programme

These are not the only variables, however, which influence the satisfaction obtained by a child from work with a teaching machine. To give him most satisfaction the programme needs to be suitable, graded appropriately in difficulty of the items, and avoid problems such as those created by spatial handicaps of cerebral palsy. Not all these criteria were met by the programme used in this study, and it would be fair to comment that results from the teaching machine programme might have been considerably more impressive if some of the children had been allowed more practice with less steeply graded items. If this were the case, satisfaction with the work should also have been even more emphatic.

The Machine

Among the drawbacks in presenting instruction in a programmed form are the practical difficulties of ensuring a robust trouble free machine. Machines for cerebral palsied children are required to withstand uncontrollable shaking of the hands, slumping of the child on to the device, excessive or prolonged pressure on various parts, and need to work easily. Whilst some modifications to the machine, mainly for the severe athetoids, were carried out, all the desirable features could not be incorporated into the design of the machine. Two weaknesses of particular inconvenience to the children were the erratic conduct of the card selector and the occasional lack of synchronisation between the frames and answers, due to faulty positioning of the cards. A specific expectation, that children suffering from multiple handicaps would benefit especially from the teaching machine approach, appeared as a trend but could not be confirmed by statistically significant differences between experimental and control groups. Once more, the teaching machine was as successful as conventional teaching, but under more favourable circumstances qualitatively better results would be expected.

PREDICTORS OF ARITHMETIC ATTAINMENT

Bender Gestalt Test and Arithmetic Attainment

As the results show, the Bender Gestalt test and the CST had major successes in predicting achievement on the arithmetic attainment tests in the study. The reasons for this are explored a little more closely and compared where appropriate with other findings.

In the review of the literature, claims were made (Dunsdon, 1952; Koppitz, 1958, 1960 and 1964) that the Bender Gestalt and intelligence tests were good predictors of arithmetic success. It appears that the distinction between the ability to compute accurately as apart from mathematical reasoning skills was not clearly established.

The criterion for arithmetic achievement used by Dunsdon was the Burt Mental Arithmetic test (1933) and Koppitz employed the WISC Arithmetic subtest as well as the Metropolitan Achievement tests.

Bender Gestalt and Mental Arithmetic Problems

Dunsdon found that performance on the Burt Mental Arithmetic test was most closely related to "visual gestalt maturational level" as measured by the Bender Gestalt test. She studied a highly selected group of thirty-five cerebral palsied children drawn from her own larger sample, and based her scoring on the mental age norms provided in the manual.

Dunsdon found that 60 percent of the children showed a closer relationship between their Bender and Arithmetic scores, as compared with 46 percent that showed similar relationships between arithmetic and verbal reasoning and visual memory factors (based on various items of the Stanford Binet 1937).

In a group of ninety elementary school children (6 yr 7 months to 11 yr 7 months) comparable to British Child Guidance referrals, Koppitz (1958) found the WISC Arithmetic subtest closely related to the Bender test. On her scoring system a higher correlation was found between the Bender scores and WISC Arithmetic subtest among older children.

In the present study, no significant relationship was noted between the WISC Arithmetic or Schonell subtest 12 (which consists of mental arithmetic problems) and the total Bender Gestalt score. The correlations between the total Bender Gestalt score and WISC Arithmetic and Schonell subtest 12 are 0.27 and 0.08 respectively.

Bender Gestalt Test and Simple Calculating Skills

Koppitz (1960, 1964) further suggested in two studies that efficiency in simple calculating skills was related to good Bender Gestalt performance. In the first investigation the Metropolitan Achievement test was administered at the end of the first and second grades, and the Bender Gestalt test was taken by the children on entering their first and second grade classes at school. She claimed that the total Bender Gestalt error score was highly consistently related to arithmetic attainment (see Table XXIV).

TABLE XXIV
CORRELATIONS BETWEEN BENDER AND ARITHMETIC
ACHIEVEMENT TEST SCORES (Koppitz 1960)

1st Grand Bender and 1st Grade Achievement	(N = 145)	—.57
2nd Grade Bender and 2nd Grade Achievement	(N = 141)	—.75*
1st Grade Bender and 2nd Grade Achievement	(N = 88)	—.49*
2nd Grade Bender and 2nd Grade Achievement	(N = 73)	—.46*

All correlations are negative since Bender is scored for errors.
* Correlation is significant at .001 level.

In the second study Koppitz compared the Bender Gestalt drawings of two groups of children with the Metropolitan Achievement Test. Children in Group 1 were administered the Bender Gestalt test at the beginning of the first grade and tested at the end of their third grade on the Arithmetic test, whilst the children in Group 2 took their Bender Gestalt test at the beginning of their second grade and were tested after completing their second grade.

70   *Arithmetical Disabilities in Cerebral Palsied Children*

Her main observations were that while no single item on the Bender Gestalt exclusively predicted success on the Metropolitan Achievement test the total error score as well as twenty-two out of the thirty individual items were significantly related to it. Unfortunately the significance of her findings is obscured because the IQ levels of the children were not indicated. Also these results are not strictly comparable as they refer to noncerebral palsied children (see Table XXV).

TABLE XXV

RELATIONSHIP BETWEEN THE BENDER GESTALT TESTS AND ACHIEVEMENT IN ARITHMETIC (Koppitz 1964)

	Arithmetic Achievement	Good Bender	Poor Bender	X^2		P
Group 1	High Arithmetic	2	4	10.64	<	01
(N = 29)	Low Arithmetic	1	12			
Group 2	High Arithmetic	23	6	22.00	<	.001
(N = 51)	Low Arithmetic	2	20			

In the present study the correlations between the Bender Gestalt total score and the arithmetic criterion tests assessing simple calculating skills proved very low.

Intelligence and Arithmetic Attainment

Phillips and White (1964) found the Stanford Binet test a poor predictor of arithmetic attainment in groups of physically handicapped children carefully matched except for cause of motor handicap.

TABLE XXVI

CORRELATIONS BETWEEN TOTAL BENDER GESTALT SCORE AND ARITHMETIC CRITERIA

	Southend	Schonell (1-5)	Schonell (8-11)
Bender Gestalt Total Score	.27	.09	.09

One group had motor handicap resulting from cerebral palsy and the other group resembled the first in motor handicap, but differed in having no signs of brain injury. The criterion for arithmetic success was the Southend Arithmetic test and the

Discussion

intercorrelations of IQ, CA and arithmetic performance are given in the table below.

In the present study the WISC yielded even poorer correlations with the arithmetic criteria.

TABLE XXVII
PRODUCT MOMENT CORRELATIONS OF CRITERION AND
PREDICTION VARIABLES (Phillips and White 1964)

	IQ	Southend Arithmetic
CA	.40	.34
Southend Arithmetic	.49	

TABLE XXVIII
CORRELATIONS OF WISC WITH ARITHMETIC CRITERIA

Variables		Southend	Schonell 1-5	Schonell 8-11
12	CST IQ	.47	.35	.47
9	WISC Full IQ	.18	.05	.15
10	WISC Verbal IQ	.15	.10	.15
11	WISC Perf IQ	.15	—.02	.10
13	DAM IQ	.10	—.06	.05

Individual items of the Bender Gestalt test as well as the WISC arithmetic subtest were significantly associated with arithmetic success. While CA correlated highly with arithmetic success ($p > .05$) this is not unexpected as improvements in skills are related to length of experiences and opportunities to do arithmetic. Children suffering from diplegia were significantly poorer ($p < .05$) at arithmetic, but no special significance is attached to this finding in view of the disagreements in classification. Nonetheless it offers some support to the speculation mentioned in the review, that children restricted in movements enjoy reduced sensorimotor experiences. This in turn impairs the acquisition of appropriate spatial concepts necessary for the basic skills in simple computation. It must be acknowledged that firmer empirical evidence would be required to support such a hypothesis.

CST

As mentioned previously the most relevant predictor of

arithmetic attainment was the CST as shown in Table XXIX below.

TABLE XXIX
CORRELATIONS OF CST WITH ARITHMETIC ATTAINMENT TESTS

	Southend	Schonell 1-5	Schonell 8-11	Schonell Total
CST	.49*	.35†	.47*	.41*

* Sig. at 1% level.
† Sig. at 5% level.

With normal subjects tests of "pure" intelligence like the CST tend to correlate highly with other measures like the WISC. The question arises as to whether the lower than expected correlations found in this study is due to the fact that this is an atypical cerebral palsy group. None the less some explanations for the success of the CST are considered and it appears that three features of the test might account for the higher level of prediction for this measure, namely intelligence, "mental work power" and memory.

Pure Intelligence. Morrisby claims that the CST is a measure of "pure" intelligence unaffected by past education and cultural stimulation and that it is uncontaminated with either perceptual or spatial components whereas the WISC is a mixture of various factors.

"In a sense the CST measures the same kind of intelligence as do the WISC, Stanford Binet and many other omnibus-type tests, but has a somewhat higher level of efficiency than most of these. This is due to a large extent to the fact that it measures the relevant function directly, whereas the WISC etc. rely on the contaminating components cancelling each other out in the combining of performances from various test-tasks, and assume that the variance that remains is indicative of the hard core of "pure intelligence" (Morrisby—Personal communication).

It suggests that arithmetic requires the ability to reason in abstract symbols which is the essence of "g" of which the CST is a good test, according to Morrisby. It is argued that reasonably high scores can be obtained on the WISC or Stanford Binet through past learning or exposure to cultural or educational experiences, enabling the child to deal with a variety of concrete tasks, rather than help him develop abstract reasoning ability.

Mental Work Power. Morrisby states that the CST is a test of an aspect of intelligence which for want of a better term he calls "mental work power." The test requires persistence and concentration in the performance of an intellectual task rather than the more intuitive solution of problems of relationship such as is required by the other tests of intelligence. The child works on his own and is neither forced nor encouraged by the examiner as in the WISC or Stanford Binet, and failure would be influenced by such personality variables as restlessness and hyperactivity. "The meaning of the concept of 'mental work power' is probably best understood by actually performing the test itself under the usual testing conditions, and by considering the relationship between the results of the test and other variables such as school marks, etc."

Memory. On introspection the writer realised how in trying to solve the problems one required to hold sequences in the head and was dependent upon immediate short-term memory. It seems obvious that such abilities are involved in learning arithmetic.

Spatial Disorders

Whilst the CST may as Morrisby claims be "functionally free of perceptual and spatial ability" in normals this is not therefore necessarily true for cerebral palsied children. The importance of perceptual and spatial difficulties is further considered particularly with regard to reproducing the Bender Gestalt figures.

The present findings on spatial disorders are in agreement with those of Dunsdon and Koppitz, as the children who were poor at integrating the BG figures or who "perseverated" tended to be weak in arithmetic. Dunsdon remarked that "the spatial sense appears to play a large part in the process of learning to read and write and it seems to enter largely into the ability to manipulate numbers." Strauss and Lehtinen observe that number concepts have their origin in perceptual experiences, and a number scheme grows out of the ability to manipulate objects in space; evolving in a "semi-abstract way from an understanding of relations, of parts to parts and parts to wholes."

Any further development and analysis of this argument depends upon the splitting up of the visuomotor functioning into three stages: (a) perceptual, (b) reasoning and (c) motor. In attempting to copy a Bender Gestalt design the child must first perceive the object correctly, next reason about its make-up and finally, at the motor stage, the brain must issue instructions to the hand to reproduce the design. Thus in Figure 6 a child must initially perceive two sets of sinusoidal curves intersecting at an angle, next reason about the composition of the loops set at certain angles and finally execute a series of motor movements to copy the original design. A child may have no perceptual difficulty and be aware that his own reproduction is an unsatisfactory likeness of the model. Anyone who has worked with cerebral palsied children will be familiar with the remark "why can't my hand do what my eyes see?"

What is required are more refined measures to assess separately these three aspects, since the Bender Gestalt and WISC (Block Design and Object Assembly subtests) are dependent on all three abilities. These need to be separated in the way Frostig's Developmental Test of Visual Perception (Frostig 1964) distinguishes (a) eye motor coordination, (b) figure ground, (c) form constancy, (d) position in space, and (e) spatial relations.

Birch and Lefford (1964) have shown how cerebral palsied children may be as good as normals in recognising and perceiving shapes but slightly weaker at analysing the shapes into their constituent parts, but find it quite impossible to reason out how the forms can be synthesised or assembled from separate elements.

It is evident that much of Piaget's contribution is relevant to our further understanding of the problem. But much sustained effort and ingenuity will be required to integrate Piaget's ideas with the attempt to distinguish various aspects of spatial functioning.

Unfortunately there appears to be a tendency to be interested in spatial difficulties for their own sake and there is no attempt in any of the researches reviewed by Abercrombie (1964) to relate spatial difficulties to educational success or failure, or developmental theories of intellectual development. We are left with a knowledge that there are connections; case studies like Caldwell's (1956) describe attempts to remedy a child's educational failures caused by "spatial inability."

In conclusion the most appropriate and satisfactory explanation for the significant correlations between the Bender Gestalt errors and success in arithmetic and in part for the significant correlations between the CST IQ and arithmetic attainment would be that they both arise from common causes, or combinations of the effects of brain injury and lack of visuomotor and perceptual experiences or an inability to assimilate them.

PART THREE

CHAPTER V

IMPLICATIONS FOR TEACHERS

W HILE THE MAJOR purpose of this investigation was to evaluate the effectiveness of programmed instruction for teaching arithmetic to cerebral palsied children, a study of this kind does throw up a number of points of much wider significance for the education of such children. These include the relationship between research which has a direct or an indirect bearing on the education of cerebral palsied children and its actual application in the classroom: the difficulties involved in evaluation of previous research; the practical and administrative problems raised by the introduction of programmed instruction; the role and problems of the educational researcher in the schools; a number of questions relating to the problem of assessment of cerebral palsied children; the sociological repercussions which must be recognised as likely to occur when new knowledge gained from research is actually introduced into the (special) schools; and finally the necessity for the staff to develop an integrated approach to the education and treatment of the cerebral palsied child and a critical approach to experiment and innovation.

RESEARCH

Lack of Theoretical Models

Review of the literature reveals few studies which specifically examine the incidence, nature and causes of arithmetical difficulties in CP children and shows that these are rarely related to psychological theories. There do appear to be three major areas of research with particular bearing on the difficulties of CP children, or on the type of approach which is needed by the teacher to overcome such difficulties. Firstly, a few psychologists do attempt to relate in a systematic way impaired functioning of the brain with arithmetical difficulties, although little use has as yet been made of their findings in the classroom. Secondly, Piaget's work may have much greater application to the teaching of CP children than has hitherto been realised. Finally, research in learning theory, especially on operant conditioning lines, could be more used in the teaching of arithmetic to CP children.

As regards the first group, the early work of Strauss and Werner provides a notable exception to the frequent failure noted above of many workers to relate difficulties and delays in the acquisition of academic skills to psychological theories. They attributed the poor academic skills in brain injured children (including those with CP) to disordered figure-ground perception, distractability, perseveration and other psychological characteristics described by the Gestalt psychologists. They tried to analyse the difficulties of these children in a systematic manner, looking at the visuomotor, tactuomotor and audiomotor skills and declared that impairment in these areas arose not only from corresponding sensory defects but could also be traced to pathological central processes. The authors concluded that weaknesses in the perceptual integration and organisation of the child were often reflected in delayed or arrested development of academic skills including ability in arithmetic.

Neither psychologists nor teachers, however, have systematically followed up these leads in the classroom situation, which seems surprising in view of the significance attached by clinicians to the effects of brain damage on perceptual and other cognitive skills. Again, educators have failed to explore the implications of such workers as Gerstmann and Kinsbourne and Warrington who attempted to relate arithmetic difficulties to neurological abnormalities. The former put forward a theory in which "acalculia" was recognised as a syndrome of cerebral cortical disorder, while Kinsbourne and Warrington attempted to relate arithmetic difficulties to impaired sequential ordering and discrimination of the fingers.

Piaget has analysed and described the developmental stages through which children progress, and the kinds of knowledge they possess at each of the stages. Hunt (1961) demonstrates the potential use of this approach and draws attention to links with animal learning studies and attempts to simulate "thinking" on a computer. Uzgiris and Hunt (1971) are developing an instrument for assessment of cognitive development on Piagetian lines. The use of systematic observational techniques to discover how normal children learn, the kinds of errors they make and how they overcome them, has been very little exploited for the benefit of cerebral palsied children. Woodward (1961) has used Piaget's approach for examining the stage of development of number concepts with subnormals and she has pointed to its diagnostic and educational value for cerebral palsied children. Teachers of cerebral palsied children need sure knowledge of the normal and deviant patterns of child development and they require instructional skills to promote the appropriate level of children's learning. Again no longitudinal or case studies have been carried out to pinpoint the great variety of individual developmental patterns in cerebral palsied children in Piagetian terms. It is possible that the development of similar sequences of skills and knowledge would be uncovered in cerebral palsied children and that the teacher using an analytic approach might effectively teach children with, for example, perceptual difficulties, auditory weaknesses or visuomotor impairment, by a series of remedial exercises. While Piaget's ideas have stimulated much thought and research, no standardized tests are as yet available to assess the level achieved by cerebral palsied children in the "preconceptual" and "intuitive" stages.

Skinner has formulated a clear application of the reinforcement technique for learning in the classroom. His methods setting out the range and availability of reinforcement techniques present in the classroom have been tested with normal children. Automatic devices to minimise the erratic control or reinforcement by teachers have not been received enthusiastically by most of them, although a few are convinced that this would free them from repetitive chores in the classroom and allow them more time with individual children.

The notion that instructional technology is analysable, that "negative information" is unhelpful, that a learner can only handle information that can be adequately processed, that various features in the environment need structuring to maximize learning has been accepted and tried out in other spheres, but not in the teaching of cerebral palsied children. Teachers need rather to discover the optimal conditions for efficient learning than to describe its features.

Some teachers have, however, been guided by learning theories. Of significance is the work of Clarke and Clarke (1965) who has offered a set of principles emphasising the use of incentives, the breakdown of the task, the spacing of learning etc., which can be used to train the severely subnormal. Also very suggestive is the body of data presented by O'Connor and Hermelin (1963) which points at "deficits in acquisition rather than to poor perception, retention or transfer ability" as a major determinant of backwardness in speech and thought among the severely subnormal.

In conclusion it is suggested that a combination of the developmental approach of Piaget and a learning theory model of Skinner could be of considerable value to the teacher. An assessment of the developmental level reached by a child, including a systematic analysis of the particular areas in which his functioning was delayed or arrested would have to precede the formulation of any teaching programme in arithmetic.

Difficulty in Comparing Studies

One of the significant factors emerging from some of the studies reviewed in the literature is that not only have these shown a lack of detailed information, but that the task of evaluating the results has been complicated by the use in many of them of confused terminology, contradictory diagnostic criteria and poor definition of variables which do not permit replication in future studies. The early endogenous-exogenous typology proposed by Strauss, for instance, is rarely adopted since the rationale by which brain injury is hypothesised from the evidence of behavioural symptoms alone is unacceptable, nor is it easy to compare some of the early studies since there is no uniform classification of the cerebral palsied types.

Another difficulty is the confusing manner in which various groups of children who are slow in learning or motor development are categorised as "brain damaged" despite the lack of definite evidence of neurological impairment in the so-called "minimal brain damaged" or minimal cerebral dysfunctioning (mcd) child. The weakness of "vague and oversimplified classification" for psychological investigations, and the tendency to include children with recognised pathologies of brain and behaviour as brain injured occurs even with Cruickshank who labels the hyperactive child as brain injured.

Most studies have confirmed that the mean intelligence quotient of cerebral palsied children is lower than that of normal children. What is striking in many of these studies is the paucity of information about the precise tests used, or which items of the Stanford Binet test, or how much allowance made for children with severe motor impairments, or unintelligible speech. Another feature of research in this area is the lack of awareness, despite documented evidence, that the widely used Stanford Binet test is inappropriate for cerebral palsied children, and that the concept of the mental age based on the Binet result is unhelpful (Clarke 1966).

Abercrombie has drawn attention to the lack of clear definition of the terms and tests used in several studies, to the variety of descriptions given to these perceptual, spatial, visuomotor and other disorders of space perception in cerebral palsied children and to a range of tests used to assess these skills. It is clear that until recently the distinction between the perceptual and visuomotor or constructional difficulties in such children was not clearly made, so that in this area too it is difficult to evaluate or compare the results of earlier studies.

Programmed Instruction and Teaching Machines

In the first flush of excitement misguided enthusiasts for programmed instruction made extravagant claims for the new technology of teaching. The undue emphasis on machines and gadgets distracted attention from the basic need for good programmes. Unfortunately the commercial exploitation of this "first really new tool of education since the printed book" has proved a serious embarrassment to educationalists, particularly in America (Edling et al., 1964). Teachers became anxious that their roles would be usurped and their "incompetence exposed" by teaching machines. When conflicting reports of the superiority of programmed instruction over their own teaching methods reached them, it added to their confusion.

The difficulties of assessing the effectiveness of programmed instruction have already been discussed in the review of the literature, and Hartley has drawn attention to the lack of sophistication in most investigations using programmed instruction as an instructional technique. Another serious problem despite the vast number of commercial programmes available, is the difficulty in identifying material suitable for the children, especially as regards mathematical programmes. Selection of programmes is complicated by the problems of distinguishing the good from the mediocre. Often the only practicable method of determining the effectiveness of such programmes is to rely on the published research evaluation. The alternative of conducting tests oneself is generally time-consuming and most schools are ill-equipped to undertake such evaluative assignments. Adaptation of programmes without such careful research not uncommonly leads to disappointment, and discourages the teacher from giving programmed instruction a further trial. In the present study, as mentioned earlier, weaknesses in the programme become apparent on examining the response rates of children, several of whom failed to obtain the minimum of 95 percent correct answers.

Teachers also justifiably argue that where programmes "guide 90 percent of the activity of students" the creative side of the teacher's activities must be "squeezed into the remaining 10 percent of the instructional programme." The teacher will have only limited time available in which to develop the children's learning readiness or to help overcome the specific difficulties encountered by individual children taking the programme. Since such programmes are unlikely to cater specifically to a group of cerebral palsied children with a wide range of individual learning difficulties, this may be a major problem for the teacher.

It would seem that programmes themselves could be divided into smaller interchangeable units which teachers could combine to meet specific needs of the children. Computer programmes, such as the Edison Responsive Environment, might also be used to provide more diversified learning opportunities. A further requirement is a degree of flexibility in the programme to meet any curricular changes. Teaching methods in the classroom can be adapted, for example, by the introduction of the "new mathematics," whereas programmed instruction may be looked on as a costly and less adaptable investment.

Programmed instruction in "certain" subjects can be used under suitable circumstances both for "enrichment" and for remedial purposes for cerebral palsied children. It could be usefully employed for children in a heterogeneous class in enabling them to proceed at their own pace and according to their own abilities. A teacher thus released part of the time from drilling children according to orthodox instructional techniques would have more time available to discover and aid individual weaknesses and problems. The successful use of programmed instruction will depend on certain basic requirements being fulfilled. These will include the cooperation of teachers and administrators participating in "the scheme" and the appropriateness of the instructional material. If these requirements are fulfilled, programmed instruction can offer an alternative form of individualised instruction, which, carried out in a favourable milieu, will ease many of the particular learning problems encountered by cerebral palsied children.

IMPLICATIONS OF THE PRESENT STUDY FOR ASSESSMENT METHODS

Use of Tests for Placement Decisions

The implications thrown up by this study for the testing of cerebral palsied children need to be carefully considered. We need to establish the accuracy of the predictive value of the IQ for their training programme. Generally the IQ has been looked upon as a reliable guide to their present intellectual functioning, and as establishing a ceiling to their future development. The validity of both these assumptions has been questioned by Sarason and Gladwin (1958) in a critical review of the literature in which they declared that "the IQ is unstable over time, cannot be reliably measured by any agreed-upon single instrument and has strong environmental determinants." The validity of the IQ placement decisions is even more questionable in the case of cerebral palsied children. Thus Phillips and White (1964) have demonstrated the failure of the Stanford Binet IQ result to predict academic attainment in cerebral palsied children.

There is clear recognition that cerebral palsied children have reduced physical movement and a lack of other experiences necessary for intellectual growth and as the predictive value of the IQ is based upon the assumption that all children have equal experiences of this kind, this assumption has a less valid basis in their case.

In this study the standardised tests of intelligence (WISC, DAM) yielded poor correlations with arithmetic attainment.

Use of Tests for Diagnostic Purposes

While tests are often used to establish the individual's present level of intellectual functioning and to make a tentative assessment of his potential, these interpretations being drawn from "concurrent" and "predictive" validity studies of the tests, they are rarely used to help teachers in the classroom. Much more useful diagnostic test batteries which guide the teachers to establish the precise assets and limitations of a child can enable them to choose the appropriate teaching methods. Without such assessments the teacher may have to rely on a "let's try everything" approach, on his or her personal preferences or even on the advertising claims for the latest educational gimmick. In the case of an experienced teacher this may still work out well in the long run, but it leaves the novice floundering.

The Illinois Psycholinguistic Battery (Kirk and McCarthy, 1961) separately assesses nine determinants of the effective use of language, named by them decoding visual and auditory stimuli, auditory-vocal and visual-motor association, vocal and motor encoding, auditory-vocal and visual-motor sequencing and auditory-vocal automatic ability. In subsequent reports the authors and others have argued for the value of the multidimensional analysis.

Many tests have been developed to assess "spatial" difficulties (Abercrombie, 1964). One important battery of particular relevance to the teaching of arithmetic would appear to be the Frostig Developmental Test of Visual Perception (Frostig, 1964). This battery claims to assess five "spatial" abilities: figure-ground discrimination and form constancy, eye-motor coordination, position in space and spatial relations. It is said that these findings help in planning teaching to overcome or circumvent the specific deficits identified.

The above batteries may be inappropriate for younger children but batteries with a similar rationale can be developed. The work of Madame Albitreccia in Paris (Borne, 1965) offers many fruitful suggestions for the contents of an assessment battery at a lower level and for possible methods of overcoming the developmental lags and defects. These tests are intended to provide a profile of children's strengths and weaknesses, and offer practical suggestions for the training of abilities, and should form a basis for planning the curriculum and selecting the teaching methods of the remedial programme.

TEACHING

Social and Organisational Implications of the Introduction of New Techniques

There are certain recurrent questions in special education which centre around the special methods of teaching physically handicapped children. During the past ten years there has been a marked increase in the application of programmed learning techniques in the classroom.

This is an instance in fact where practical questions as well as theoretical issues are encountered. These are practical questions, because it needs to be established whether the programmed techniques of a type currently offered for use are appropriate for cerebral palsied children or for that matter acceptable to teachers. However good the method might be under ideal conditions, it can only reach the child after it has been accepted by the teacher.

It is also a theoretical issue whether the social roles, attitudes and prejudices that teachers carry to their work can be recognised in their reactions to a novel approach to teaching.

The new knowledge brought by research findings will also have sociological repercussions, since educators, administrators, inspectors of schools and others concerned with the education of handicapped children now need to examine their own status and role in the present social organisation of physically handicapped children. In the past many of the assumptions about the care and education of such children have been taken for granted and changes were seldom entertained. But today, with the rapid spread of new knowledge and the introduction of new techniques, professionals are subjected to all kinds of strains and stresses. Teachers especially, as society's representatives of traditional values, find themselves in conflict with newer developments. They are unsure where they should be resisting change and where they should be welcoming it. When educationalists, administrators and doctors are required to reorientate their thinking and themselves to initiate social changes, they need sympathetic support if they are not to be unduly upset.

In traditional work on teaching the general accent is on the child's response to the teaching method, teacher and other children. Where psychologists have investigated teachers, their studies have tended to be concerned with attitudes, role, status etc. rather than the immediate incentives obtained from the teacher's work. The introduction of teaching machines will undoubtedly change to some extent the roles of both student and teacher. The former will interact more directly with the material and in a more independent way, while the role and task of the teacher will be altered in two major areas. Firstly, he will give up some of the routine part of his work such as imparting practical information and teaching and testing of the simple skills, more emphasis now being placed on his responsibility and opportunity for personal guidance and motivation of the pupil, individualising his teaching and stimulating critical and creative thinking. Secondly, as Jaspers (1967) points out, his attitude too will have to change and become more "critical and scientific," particularly as regards the specification of his purposes.

As he will depend more on the producer outside the class, he will function less as an authority and more as a sharer in the instructional process.

Less frequently recognised is the fact that the use of teaching machines immediately cuts across the incentive obtained by teachers. To the extent that incentives are reinforced by the rate and substance of a reward this is bound to have a considerable effect on teaching and the teacher. Interposing a machine between a child and the teacher delays the rewards inordinately for the teacher and the effects of this are well documented. It is not therefore surprising that many teachers resent the introduction of teaching machines, even though they are not aware of it. This is not to suggest that this is the only reason for the teacher's lack of enthusiasm. Another reason is that within the interpersonal teaching situation the direction of work is under constant control by the teacher who modifies and directs the rate of the responses of the child, whereas with the teaching machine this is not so or at least not obviously the case. If a standardised pattern of work has been laid down in advance and there is none of the continuing reinforcement for the teacher that the steps on the programme are the best and correct steps to be pursued in teaching, uncertainty will develop which in time leads to insecurity. Furthermore, the personal involvement which rests on so many aspects of rewards and which is built up from the mutually reinforcing dialogue between teacher and pupil and responses between a teacher and his pupil quite outside the immediate scope of the curriculum needs is weakened. In other words, the nonacademic rewards of the interpersonal relationships are felt to be reduced in proportion to the place occupied by teaching machines.

As regards the teacher's general attitude towards research, a surprisingly naïve assumption prevails that teachers ought to share the enthusiasm of the research worker for experiment and innovation in education. It is often forgotten that the teacher needs to be consulted when changes in the educational organisation are being considered. Unless teachers are reasonably reassured that their natural style of instruction is basically in harmony with any proposed technological changes, they are likely to resist change. For example, should teachers feel that their children's experiences would be limited by programmed instruction and that social skills and interpersonal relationships might become restricted by it, they would consciously or unconsciously withhold their support for its introduction.

Teachers in special schools are more often targets for attention by social agencies, medical and auxiliary staff than those in ordinary schools. A more effective defence is built up by the teachers against interference and often well-meant suggestions, to safeguard their own methods of work, as well as their relationship to their children. For this reason research of this type is best carried out through the agency of the teacher. In this way the information is not restricted to the performance of the children and the results of tests, but it also offers scope for a wide observation of practical responses by teachers and allows attempts to evaluate their influence on the way in which teaching in special schools is performed in practice.

Without a sure knowledge of the children they teach and the feeling that they have something unique to offer, teachers will remain insecure when communicating with other specialist colleagues such as the doctor or psychologist and it seems that interprofessional respect will remain an empty platitude.

When a research project is being undertaken or new techniques introduced, success will depend to a large extent on the development by the members of staff of an integrated approach. Where this does not exist, curriculum planning cannot be fully successful. This is especially true if the introduction of programmed instruction means that administrative procedures, such as timetabling and grouping by students, will be radically altered. Naturally each specialist, whether a teacher, therapist, houseparent, psychologist or doctor, brings a different approach and outlook and has something unique to offer, but they must be willing to learn from one another how to integrate their contributions. Not only must the approach be integrated in purpose, but it must also be experienced by the child as an integrated relationship with the staff. Teachers, houseparents and therapists may each teach and treat him in a different way, so that the child will feel fragmented. To develop a unified self-image he needs consistency as well as warm and emotionally satisfying relationships. This would be promoted by formal and informal exchanges of opinions and information among the staff. At these meetings research needs and methods could be established, the ways in which they would be likely to impinge on each of the specialists in the schools discussed and the results of research at each stage communicated.

SUMMARY

1. The aims of the study were:
   a. To review the factors which make learning arithmetic difficult for cerebral palsied children.
   b. To undertake an exploratory study to see whether
      (1) CP children could be taught arithmetic through Programmed Instruction.
      (2) they could learn as effectively through Programmed Instruction as through conventional methods.
      (3) Programmed Instruction was more effective than conventional methods.
2. The experimental design was as follows:
   a. Selection of subjects.
   These were drawn from three schools for cerebral palsied children, two being secondary modern and one bilateral. In each school two entire classes took part in the experiment, each class being divided almost equally into an experimental and a control group, the groups being matched as far as possible for handicap, age and sex. The same teacher spent an equal amount of time with both the control and experimental group. A total of forty-two children between the ages of nine and sixteen took part. Both sexes were included and there was a considerable range of handicaps.
   b. Collection of data prior to administration of the programme.
   Since the cerebral palsied child's ability to learn depends on an even wider range of factors than does the normal child's, it was considered important to obtain for each subject data on (a) physical disabilities, particularly sensory and visuoperceptual disorders; (b) intellectual ability; (c) emotional factors, (d) social background.

Summary

The medical and social data were obtained largely from records, and intellectual and emotional data from a series of tests administered to each subject over a period of weeks before the experimental programme. These included (a) The Compound Series Test, (b) The Wechsler Intelligence Scale for Children, (c) The Goodenough Draw-a-Man Test, (d) The Bender Visual Motor Gestalt Test, (e) The Bristol Social Adjustment Guide.

c. Administration of the arithmetic programme.

The programme of instruction, which was applied over a period of thirteen weeks, covered the four basic rules of arithmetic. The control groups were taught by conventional methods and the experimental groups through the ESA Tutor machine using the "Primary Arithmetic" programme written by R. D. Bews.

d. Measurement of arithmetic attainment.

The tests used to observe differences in postlearning attainment between the experimental and control groups were:

(1) The Southend Attainment Test in Mechanical Arithmetic.

(2) The Schonell Diagnostic Arithmetic Tests.

3. Comparison of the experimental and control groups showed the following results:

a. Using a base-line established by increases in arithmetic attainment showed that learning had taken place in both groups. It was firmly established that cerebral palsied children are able to manipulate a teaching machine and that their progress is sufficiently rapid to make this method of instruction of practical classroom use. The results suggested that the anxiety expressed by several teachers about the value of programmed instruction were unfounded, and that, particularly in the light of the shortage of available staff, programmed instruction has an important role to play.

b. No significant differences were found in the overall level of attainment of children in the two groups.

c. Such differences as there were, however, were in favour of programmed instruction as the method, for the following reasons:
(1) The overall time spent with the teaching machine was less than that offered by standard instruction.
(2) There was some evidence that programmed instruction was particularly helpful to children with interpersonal difficulties, including those who were restless or withdrawn.
(3) The standard deviation of the means and the scatter of individual attainment levels in the control group was nearly three times greater than for children in the experimental group, children taught by programmed instruction showing far greater consistency in their level of attainment. This suggests that more homogeneous results can be expected from programmed instruction than from conventional teaching methods.
(4) Certain factors clearly militated against the success of the experimental group. These factors are not inherent in programmed instruction, and once removed it is reasonable to suggest that programmed instruction might show up as significantly superior to conventional methods. They included the fears of teachers that programmed instruction posed a threat to them and the unfamiliarity of the children with this method of instruction coupled with certain practical difficulties experienced in manipulating the machines. The programme had not been designed specifically for cerebral palsied children and did not take into account, for example, their spatial difficulties. A more refined series of stages is needed since the cerebral palsied child requires a more careful grading of intellectual demands than the normal child. The base line of the programme must also be much more closely related to the age of development and level of intelligence of the child.

d. Prediction of the progress of arithmetical attainment was found to correlate with the spatial and nonverbal items of tests.
4. Finally, several general points emerge from this study.
   a. Skilled diagnostic testing procedures (such as the Frostig Test and ITPA) are urgently needed for identifying cerebral palsied childrens' difficulties more precisely.
   b. Teachers of the physically handicapped child lack theoretical bases. One approach could be to relate the findings of Piaget and Skinner that is, first the stage of development should be identified and then the appropriate programme offered.
   c. There is a great need for cooperation between research workers, psychologists and teachers in the devising of systematic programmes and the evaluation of different teaching methods (such as Frostig) for use with cerebral palsied children. This exploratory study has suggested that programmed instruction in arithmetic carries many advantages for cerebral palsied children and it is one of the many areas in which research would be valuable.

BIBLIOGRAPHY

Abercrombie, M. L. J.: Perception and eye movements; some speculations on disorders in cerebral palsy. *Cerebr Palsy Bull,* 2:142-148, 1960.
Abercrombie, M. L. J.: Eye movements, perception and learning. In Smith, V. H.: Visual Disorders in Cerebral Palsy. London, Spastics Society/Heinemann, 1963.
Abercrombie, M. L. J.; Davis, J. R., and Shackel, B.: Pilot study of version movements of eyes in cerebral palsied and other children. *Vision Res,* 3:135-153, 1963.
Abercrombie, M. L. J.: *Perceptual and Visuomotor Disorders in Cerebral Palsy.* Clinics in Develop. Med. 11. London, Heinemann, 1964.
Abercrombie, M. L. J.; Gardiner, P. A.; Hansen, E.; Jonckheere, J.; Lindon, R. L.; Solomon, G., and Tyson, M. C.: Visual, perceptual and visuomotor impairment in physically handicapped children. *Percept Mot Skills,* 18:561-625, 1964.
Barakat, M. K.: Factors underlying the mathematical abilities of grammar school pupils. *Br J Educ Psychol,* 21:239-240, 1951.
Bax, M. C.: Terminology and classification of cerebral palsy. *Dev Med Child Neurol,* 6:295-297, 1964.
Bender, L.: *The Visual Motor Gestalt Test and Its Clinical Use.* New York, Am Ortho-psychiat Assn, 1938.
Bensberg, G. J.: The relationship of academic achievement of mental defectives to mental age, institutionalization and etiology. *Am J Ment Defic,* 58:327-330, 1953.
Benton, A. L.; Hutcheon, J. F., and Seymour, E.: Arithmetic ability, finger-localization capacity and right-left hand discrimination in normal and defective children. *Am J Orthopsychiatry,* 21:756-766, 1951.
Beskow, B.: Mental disturbance in premature children at school age. *Acta Paediat,* 37(Suppl):125, 1949.
Biggs, J. B.: The development of number concepts in young children. *Educ Res,* 1(No. 2):17-34, 1959a.
Biggs, J. B.: Attitudes to arithmetic—Number anxiety. *Educ Res,* 1(No. 3):6-21, 1959b.
Biggs, J. B.: 62nd Year Book, 1963.
Birnbrauer, J. S.: *The Ranier School Programmed Learning Classroom.* Seattle, Univ. of Washington, 1962.
Birnbrauer, J. S.; Bijou, S. W.; Wolf, M. M.; Kidder, J. D., and Tague, C. M.: *A Programmed Instruction Classroom for Educable Retardates.* (Mimeographed report) Seattle, University of Washington, 1964.

Birch, H. G., and Lefford, A.: Two strategies for studying perception in "brain-damaged" children. In Birch, H. G. (Ed.): *Brain Damage in Children*. New York, Williams & Wilkins, 1964.
Borne, G. le: A training method for reducing perceptual difficulties. *Spastics News*, 11 (No. 9):12-13, 1965.
Brownell, W. A.: *Arithmetic in Grades I and II—A Critical Summary of New and Previously Reported Research*. Durham, North Carolina, Duke Univ. Press, 1941.
Bruner, J. S.: In Helder and Piaget: The growth of logical thinking. I—A psychologist's viewpoint. *Br J Psychol*, 50:363-71, 1959.
Burt, C.: *Mental and Scholastic Tests*, 2nd ed. London, King & Son, 1933.
Burt, C.: *The Backward Child*. London, Univ. of Lond. Press, 1937.
Buswell, G. T.: A comparison of achievement in arithmetic in England and Central California. *Arithmetic Teacher*, 4:1-9, 1958.
Caldwell, E. M.: *A Case of Spatial Inability in a Cerebral Palsied Child*. London, British Council for the Welfare of Spastics, 1956.
Capobianco, R. J.: Quantitative and qualitative analyses of endogenous and exogenous boys on arithmetic achievement. *Monogr Soc Res Child Dev*, 19:101-141, 1956.
Cartwright, G. P.: *Two Types of Programmed Instruction for Mentally Retarded Adolescents*. Unpub. Masters Thesis. Urbana, Univ. of Illinois, 1962.
Cheney, A. P.: *Evaluation of Elementary School Teachers' Understanding of Basic Arithmetic Concepts*. Unpub. Master of Arts Thesis. Los Angeles, University of Califorinia, 1961.
Churchill, Eileen E.: *Counting and Measuring*. Toronto, University of Toronto Press, 1961.
Clarke, A. D. B.: *Recent Advances in the Study of Subnormality*. London, Nat. Assn. for Mental Health, 1966.
Clarke and Clarke: *Mental Deficiency; The Changing Outook*, 2nd ed. London, Methuen, 1965.
Cruickshank, W. M.: Arithmetic ability of mentally retarded children. I. Ability to differentiate extraneous materials from needed arithmetic facts. *J Educ Res*, 42:161-170, 1948.
Cruickshank, W. M.: Arithmetic ability of mentally retarded children. II. Understanding arithmetical processes. *J Educ Res*, 42:279-288, 1948a.
Cruickshank, W. M.: Arithmetic work habits of mentally retarded boys. *Am J Ment Defic*, 52:318-330, 1948b.
Cruickshank, W. M., and Raus, G. M. (Eds.): *Cerebral Palsy: Its Individual and Community Problems*. New York, Syracuse Univ. Press, 1955.
Cruickshank, W. M.; Bice, H. V., and Wallen, N. E.: *Perception and Cerebral Palsy*. New York, Syracuse University Press, 1957.
Dienes, Z. P.: The growth of mathematical concepts in children through experience. *Educ Res*, 2(No. 1):9-28, 1959.

Dolphin, J. E., and Cruickshank, W. M.: Tactual motor perception of children with cerebral palsy. *J Personality*, 20:466-471, 1952.

Drillien, C. M.: *The Growth and Development of the Prematurely Born Infant*. Edinburgh and London, Livingstone, 1964.

Dunsdon, M. I.: *The Educability of Cerebral Palsied Children*. London, Newnes Educational Publ. Co., 1952.

Dutton, W. H.: University students' comprehension of arithmetical concepts. *Arithmetic Teacher*, 8:60-64, 1961.

Dutton, W. H.: *Evaluating Pupils' Understanding of Arithmetic*. Englewood Cliffs, N.J., Prentice Hall, 1964.

Edling, J. V.; Foshay, A. W.; Ginther, J. R.; Schramm, W., and Thelen, H.: *Four Case Studies of Programmed Instruction*. New York, Fund for the Advancement of Education, 1964.

Feldhusen, J. K., and Klausmeier, H. J.: Anxiety, intelligence and achievement in children of low, average and high intelligence. *Child Dev*, 33:403-409, 1962.

Ferster, C. B., and Sapon, S. M.: An application of recent developments in psychology to the teaching of German. *Harvard Educ Rev*, 28:59-69, 1958.

Floyer, E. B.: *A Psychological Study of a City's Cerebral Palsied Children*. London, British Council for the Welfare of Spastics, 1955.

Frostig, M.; Lefever, D. W., and Whittlesey, J. R. B.: *The Marianne Frostig Development Test of Visual Perception*, 3rd ed. Palo Alto, Consulting Psychologists Press, 1964.

Fry, E.: Teaching machine dichotomy: Skinner versus Pressey. *Psychol Rep*, 6:11-14, 1960.

Gallagher, J. J.: *The Tutoring of Brain-Injured Mentally Retarded Children: An Experimental Study*. Springfield, Thomas, 1960.

Gerstmann, J.: Fingeragnosie: Eine umschriebene störung des Orientierung am eigenen Körper. *Wien Klin Wochenschr*, 37:1010-1032, 1924.

Glennon, V. J.: *A Study of the Growth and Mastery of Certain Basic Mathematical Understandings on Seven Educational Levels*. Unpublished Doctoral Dissertation, Harvard University, 1948.

Goodenough, F. L.: *Measurement of Intelligence by Drawings*. London, Harrap, 1926.

Gregory, R. E.: Unsettledness, maladjustment and reading failure: A village study. *Br J Educ Psychol*, 35:63-68, 1965.

Hartley, J.: Research report—a bibliography of research on programming. *New Education*, 2:29-35, 1966.

Hebb, D. O.: *The Organization of Behaviour*. New York, Wiley, 1949.

Henderson, J. L.: *Cerebral Palsy in Childhood and Adolescence*. Edinburgh and London, Livingstone, 1961.

Hopkins, T. W.; Bice, H. V., and Colton, K. C.: *Evaluation and Education*

of the Cerebral Palsied Child. New Jersey Study. Washington, Council for Exceptional Children, 1954.
Howard, P. J., and Worrell, C. H.: Premature infants in later life. Pediatrics, 9:577-584, 1952.
Hunt, J. McV.: Intelligence and Experience. New York, Ronald, 1961.
Jaspers, A. A.: Some aspects of the acceptability of programmed instruction in the school. The New Era, 10:225-228, 1967.
Kay, H.; Annett, J., and Sime, M. E.: Teaching Machines and Their Use in Industry. London, H.M.S.O., 1963.
Kephart, N. C.: The needs of teachers for specialised information on perception. In Cruickshank, W. M. (Ed.): The Teacher of Brain Injured Children. Syracuse, University Press, 1966.
Keislar, E. R.: The development of understanding in arithmetic by a teaching machine. J Educ Psychol, 50:247-253, 1959.
Kirk, S. A., and McCarthy, J. J.: The Illinois test of psycholinguistic abilities—an approach to differential diagnosis. Am J Ment Defic, 66:399-412, 1961.
Kirk, S. A.: Educating Exceptional Children. Boston, Houghton Mifflin, 1962.
Knobloch, H.; Rider, R.; Harper, P., and Pasamanick, B.: Neuro-psychiatric sequelae of prematurity. JAMA, 161(Suppl. 581): 1956.
Koppitz, E. M.: The Bender Gestalt test and learning disturbances in young children. J Clin Psychol, 14:292-295, 1958.
Koppitz, E. M.: The Bender Gestalt test for children: A normative study. J Clin Psychol, 16:432-435, 1960.
Koppitz, E. M.: The Bender Gestalt Test for Young Children. New York and London, Grune & Stratton, 1964.
Leith, G. O. M.: A Handbook of Programmed Learning. Birmingham, Univ. of Birmingham, 1964.
Lovell, K.: The Growth of Basic Mathematical and Scientific Concepts in Children. London, University of London Press, 1961.
Lunzer, E. A.: A pilot study for a quantitative investigation of Jean Piaget's original work on concept formation: A footnote. Educ Rev:193-200, 1961.
Lynn, R.: Temperamental characteristics related to disparity of attainment in reading. Br J Educ Psychol, 27:62-67, 1957.
Machover, K.: Personality Projection in the Drawing of the Human Figure. Springfield, Thomas, 1949.
MacKeith, R. C.; Mackenzie, I. C. K., and Polani, P. E.: Definition of cerebral palsy. Cerebr Palsy Bull, 1:23, 1959.
Malpass, L. F.: Programmed instruction for retarded children. In Baumeister, A. A. (Ed.): Mental Retardation: Appraisal, Education and Rehabilitation. London, University of London, 1968.

Matthews, C. G., and Folk, E.: Finger localisation, intelligence and arithmetic in mentally retarded subjects. *Am J Ment Defic*, 69:107-113, 1964.
Meddleton, I. G.: *An Experimental Investigation into the Systematic Teaching of Number Combinations in Arithmetic.* Unpublished Ph.D. Thesis. University of London, 1954.
Morrisby, J. R.: *Differential Test Battery.* London, N.F.E.R., 1955.
Newman, C. J., and Krug, O.: Problems in learning arithmetic in emotionally disabled children. *J Child Psychiat*, 3:413-429, 1964.
Nielsen, H. H.: *A Psychological Study of Cerebral Palsied Children.* Munksgaard, Copenhagen, 1966.
O'Connor, N., and Hermelin, B. F.: *Speech and Thought in Severe Subnormality.* London, Pergamon, 1963.
Passamanick, B., Rogers, M. E., and Lilienfeld, A. M.: Pregnancy experience and the development of behaviour disorder in children. *Am J Psychiatry*, 112:613, 1956.
Phillips, C. J., and White, R. R.: The prediction of educational progress among cerebral palsied children. *Dev Med Child Neurol*, 6:167-174, 1964.
Piaget, J.: *The Child's Conception of Number.* London, Routledge & Kegan Paul, 1952.
Piaget, J.: How children form mathematical concepts. *Sci Am*, 189:74-79, 1953.
Piaget, J.: *The Origin of Intelligence in the Child.* London, Routledge & Kegan Paul, 1953a.
Popham, W. J.: The impact of programmed instruction on conventional instruction. In Ofiesh, G. D., and Meirhenry, W. C. (Eds.): *Trends in Programmed Instruction.* Washington, DAVI-NSPI, 1964.
Porter, D. A.: A critical review of a portion of the literature on teaching devices. *Harv Educ Rev*, 27:126-147, 1957.
Prechtl, H. F. R., and Dijkstra, J.: Neurological diagnosis of cerebral injury in the newborn. In *Prenatal Care.* Collected papers and discussion presented at the symposium held at Groningen-Rotterdam 1959, Berge, B. S. Groningen, Noordhoff, 1960.
Price, J. E.: *A Comparison of Automated Teaching Programmes with Conventional Teaching Methods as Applied to Teaching Mentally Retarded Students.* Tuscaloosa, Ala., Partlow State School and Hospital (Mimeographed), 1962.
Reid, L. L.: Children with cerebral dysfunction. In Kirk, S. A., and Weiner, B. B. (Eds.): *Behavioural Research on Exceptional Children.* The Council for Exceptional Children, Washington, N.E.A., 1963.
Rogers, M. E.; Lilienfeld, A. M., and Pasamanick, B.: *Prenatal and Paranatal Factors in the Development of Children's Behaviour Disorders.* Copenhagen, Munksgaard, 1955.
Rutter, M.: Brain damaged children. *New Education*, 3:10-13, 1967.

Sarason, S. B., and Gladwin, T.: Psychological and cultural problems in mental subnormality: A review of research. *Genet Psychol Monogr*, 57:3-290, 1958.
Schonell, F. E.: *Educating Spastic Children*. Edinburgh, Oliver & Boyd, 1956.
Schonell, F. J., and Schonell, F. E.: *Diagnosis and Remedial Teaching in Arithmetic*. Edinburgh, Oliver & Boyd, 1957.
Schonell, F. J., and Schonell, F. E.: *Diagnostic and Attainment Testing*, 4th Ed. Edinburgh, Oliver & Boyd, 1960.
Skinner, B. F.: The science of learning and the art of teaching. *Harv Educ Rev*, 24:86-97, 1954.
Skinner, B. F.: *The Technology of Teaching*. New York, Appleton Century Crofts, 1968.
Shirley, M.: A behaviour syndrome characterizing prematurely born children. *Child Dev*, 10:115-128, 1939.
Silberman, H. F.: Self-teaching devices and programmed materials. *Rev Educ Res*, 32:119-193, 1962.
Smith, V. H. (Ed.): *Visual Disorders in Cerebral Palsy*. London, Spastics Society/Heinemann, 1963.
Southend: *Southend Attainment Test in Mechanical Arithmetic*. London, Harrap, 1939.
Stolurow, L. M.: Teaching machines and special education. *Educ Psychol Measur*, 20:429-448, 1960.
Stolurow, L. M.: *Teaching by Machine*. U.S. Dept. of Health, Education and Welface, Office of Education, Co-operative Research Monograph, 6. Washington, D.C., Government Printing Office, 1961.
Stolurow, L. M.: Programmed instruction for the mentally retarded. *Rev Educ Res*, 33:126-136, 1963.
Stolurow, L. M., and Walker, C. C.: A comparison of overt and covert responses in programmed learning. *J Educ Res*, 55:421-432, 1962.
Stott, D. H.: *The Social Adjustment of Children: Manual to the Bristol Social Adjustment Guide*. London, Univ. of London Press, 1963.
Strauss, A., and Werner, H.: Deficiency in the finger schema in relation to arithmetic disability (Finger agnosia and acalculia). *Am J Orthopsychiat*, 8:719-725, 1938.
Strauss, A. A., and Lehtinen, L. E.: *Psychopathology and Education of the Brain Injured Child*. New York, Cruno & Stratton, 1947.
Taylor, E. M.: *Psychological Appraisal of Children with Cerebral Defects*. Cambridge, Mass., Harvard Univ. Press, 1959.
The National Society for the Study of Education: *Child Psychology. The Sixty-Second Yearbook*. Chicago, Univ. of Chicago Press, 1963.
Uzgiris, I. C., and Hunt, J. McV.: Ordinal scales of psychological development in infancy. In Haywood, H. Carl (Ed.): *Social-Cultural Aspects of Mental Retardation*. Proc. Peabody–NIMH Conf. New York, Appleton-Century-Crofts, 1971.

VanEngen, H., and Gibb, E. G.: Structuring arithmetic. *Instruction in Arithmetic.* In *Twenty-fifth Yearbook.* Washington, D. C., The National Council of Teachers of Mathematics, 1960.

Wechsler, D.: *Wechsler Intelligence Scale for Children.* New York, Psychological Corporation, 1949.

Wedell, K.: The visual perception of cerebral palsied children. *J Child Psychol Psychiat,* 1:215-227, 1960.

Werner, H., and Garrison, D.: Measurement and development of the finger schema in mentally retarded children: Relation of arithmetic achievement to performance on the finger schema test. *J Educ Psychol,* 33:252-264, 1942.

Williams, J. D.: Teaching arithmetic by concrete analogy. I. Miming devices. *Educ Res,* 3:112-125, 1961.

Woodward, M.: Concepts of number of the mentally subnormal studied by Piaget's method. *J Child Psychol Psychiat,* 4:249-259, 1961.

APPENDIX
VARIABLES

TABLE A-I
COMPOSITION OF THE SAMPLES

No.	Group Data
1	Experimental Group
2	Age
3	Sex
4	Athetoid
5	Diplegic
6	School A
7	School B
8	School C

TABLE A-II
INTELLIGENCE TEST RESULTS

No.	IQ Variables
9	WISC Full IQ
10	WISC Verbal IQ
11	WISC Performance IQ
12	Morrisby IQ (CST)
13	Draw-a-Man IQ (DAM)
14	WISC "Verbal" factor
15	WISC "Spatial" factor
16	WISC–Information
17	WISC–Arithmetic
18	WISC–Picture Completion
19	WISC–Picture Arrangement

TABLE A-III
DISCREPANCY SCORES, I.E. DIFFERENCES BETWEEN THE SCORES ON ITEMS OF THE WISC AND DRAW-A-MAN, OR PAIRED ITEMS OF THE WISC

No.	Discrepancy Variables
20	WISC (Verbal IQ–Performance IQ)
21	WISC (Verbal factor–Spatial factor)
22	WISC (Verbal IQ–Draw-a-Man)
23	WISC Performance IQ–Draw-a-Man
24	WISC Full IQ–Draw-a-Man

104 *Arithmetical Disabilities in Cerebral Palsied Children*

TABLE A-IV
BENDER GESTALT SCORES

No.	Bender Gestalt Variables
25	Total BG Score
26	Koppitz—Distortion
27	Koppitz—Rotation
28	Koppitz—Integration total
29	Integration Fig. 4
30	Integration—3(a)
31	Integration—5(a)
32	Koppitz—Perseveration

TABLE A-V
SCORES OBTAINED FROM THE BRISTOL SOCIAL ADJUSTMENT GUIDE

No.	Bristol Social Adjustment Guide
33	Total symptoms
34	Graded symptoms
35	Unforthcoming and withdrawal
36	Depressed
37	Anxiety about adult interest
38	Hostility to adults
39	Unconcern for adults
40	Anxiety for approval
41	Hostility to children
42	Restlessness
43	Maladjustment
44	Miscellaneous nervous symptoms

TABLE A-VI
SCORES ON THE VARIOUS ARITHMETIC TESTS AND THEIR SUBTEST SCORES

No.	Arithmetic Criteria Variables
45	Southend
46	NFER Arithmetic tests
47	Schonell Total
48	Schonell Subtests 1-5
49	Schonell Subtests 8-11
50	Schonell Subtest 12
51	Schonell Subtest 3 (Multip)
52	Schonell Subtest 4 (Division)
53	Schonell Subtest 5 (Miscellaneous)
54	Schonell Subtest 6 (Graded addition)
55	Schonell Subtest 7 (Graded subtraction)
56	Schonell Subtest 8 (a Graded multip)
57	Schonell Subtest 9 (Graded division)

INDEX

Abnormalities, 19
Absence, 16, 17, 23
Abstract Symbols, 72
Academic, 23
 achievement, 27
 attainment, 86
 skills, 80
Acalculia, 80
 Gerstmann Syndrome, 10
Adding, 8
 addition, 15, 26, 45, 47, 104
Anxiety, 93, 104
 & arithmetic, 8, 16, 23
 parental, 28
 & scholastic attainments, 9
 teachers', 41-42
Apparatus
 graded, 20
 Montessori, 18
 special, 17
 structural, 14, 17, 20
 suitable teaching, 15
Arithmetic, 19, 73, 92
 ability, 16
 achievement, 12
 age, 47, 63
 c.p. child, x, 23-30, 35, 79, 80, 92
 criteria variables, 104
 criterion tests, 70
 difficulty with, 6, 7, 8, 10-11, 79, 80
 experiences in, 12, 15
 fear of, 9
 learning of, x, 11-12, 44, 92
 mental, 47, 69
 P.I., 5, 30-34, 79, 80, 95
 programme, 40, 43, 46, 93
 progress in, 17, 47
 skills, ix, 26, 35
 success in, 16, 60-63
 teaching of, x, 5, 12, 15, 19, 80, 87
 tests, 46, 70
 understanding in, 13, 20

Arithmetic Attainment, x, xi, 16, 17, 49, 52-53, 61
 & central disorders, 25
 & c.p. children, 5, 28, 30
 & brain injury, 26-27
 & disordered finger localisation, 11
 & distractibility, 30
 & emotional factors, 8-10, 28
 factors influencing, 15
 & girls, 9
 measurement of, 40, 93
 & mentally retarded, 6-8, 16
 & non c.p. children, 5, 16, 30
 & peripheral disorders, 24-25
 prediction of, 68-75, 95
 Southend Test, 46
Arithmetic Readiness, 5
 assessment of, 12
 & pre-school experiences, 17
Arithmetic Work Habits
 mentally retarded & normal children, 6, 7, 16
 qualitative & quantitative, 27
Assessment, 75
 of arithmetic attainment, x
 of arithmetic readiness, 12
 of cognitive development, 81
 of c.p. children, 79
 individual, 22
 methods, 85-87
 preliminary, 52-53
Assimilate, 75
At Risk, 28
Ataxic, xii, 23
Athetoid, xii, 44, 50, 61, 62, 67, 103
 & spastics, 20, 23
Attention-Span, 28
Attitude, 58
 of parents, 28
 of school, 39
 of teachers, 12-13, 16, 88, 89
Audiomotor Skills, 80

105

Attributes, 39
Auditory
  perceptual disorders, 20
  stimuli, 19, 86
  -vocal association, 86
  -vocal automatic ability, 86
  -vocal sequencing, 86
  weaknesses, 81

Background
  & figure, 19, 28, 30
  home, 12
Backward, 28
  backwardness, 82
Behaviour, 19, 31, 58, 83
  abnormal, 29
  analysis of, 30
  disturbance, 28
Bender Visual Motor Gestalt Test, 26, 43, 49, 51, 61, 68-70, 73, 74, 75, 93, 104
Boys
  & arithmetic attainment, 10
  premature children, 29
Brain-Injured, xi, 18, 19, 21, 28, 29
  brain injury, 17, 18, 25, 26, 27, 70, 75, 80, 82, 83
  & number operations, 10
  teaching of, 19
Branching, 32, 34, 44
Bristol Social Adjustment Guide (Stott), 43, 49, 52, 93, 104
Burt Word Reading Test, 26

Calculating
  & mentally retarded, 7
  skills, x, xi, 69-70
Cerebral Palsied Children, ix, 19, 42, 43, 68, 73, 74
  cerebral palsy, 20, 26, 39, 70, 72
  & arithmetic, 23-30
  assessment of, 79
  definition, xi
  & distractibility, 28
  education of, 79
  & emotional factors, 28-29
  & exploration of environment, 25
  in Britain & elsewhere, 22
  intelligence & motor disorders, 22
  learning difficulties of, 17-23, 79, 92, 95
  learning pace of, 56
  & normals, 20, 30
  & ocular defects, 24
  & P.I., 34-36, 65-68, 79
  schools for, 40-41
  & simple visual tasks, 24-25
  teaching of, 80, 82
  with motor handicap, 25-26
Clinical, 11
  descriptions, 28
  observations, 18
Compound Series Test (C.S.T.), 43, 71-73, 75, 93
Computational Practice, 15
  lack of, 17
Computer, 81
  analysis, 48
  programmes, 85
Concept
  of brain injury, 18
  development, 5, 14
  mathematical, 6, 7, 10, 12, 13, 14, 16, 23, 25, 30
  of "mental work power," 73
  new, 14
  of number, 14
  spatial, 71
  specific, 13
  teacher's understanding of, 12-13
  teaching of, 12, 13
Concrete
  aids, 15
  experiences, 13, 15
  materials, 12
  solution of problems, 7
  tasks, 73
Congenital
  brain injury, 25-26
Conservation
  idea of, 14
Constituent Elements
  integrating to form a whole, 30
Constructed Response, 31
Constructional Difficulties, 83
Contingencies, x

Index

Co-ordination Difficulties, 18
Correlations
  matrix, 60-63
  multiple, 26
  overall, of character qualities &
    mathematical attainments, 8
Cortical
  disorder, 80
Creative, 84
Criterion, 23, 67
  for arithmetic achievement, 68, 70,
    71, 104
  contradictory diagnostic, 82
  inadequate evaluative, 34
  minimal, 34
  variables, 49
Crutches, 44
Cultural, 9
  experiences, 72-73
  stimulation, 72
Curiosity
  environment arousing, 14

Deaf, 60
Decoding, 86
Descriptive Data, 49
Developmental
  lag, 25, 87
  level, 82
Diagnostic, 10
Diamond, 25
Diplegia, xi
  diplegic, 50, 61, 62, 71, 103
Discriminating
  discriminate, 24
  discrimination of fingers, 80
  forms, 24, 30
  tasks, 19
Distractibility, 19, 28, 30, 80
  of "brain-injured" children, 19
Dividing, 8
  division, 15, 45, 47, 104
Draw-A-Person Test, 43
Dyscalculic Children, 9

Emotional
  behaviour, 18, 19
  climate, 66
  data, 93
  demands, x
  disorder, ix
  disturbances, 8, 17, 30
  experiences, 6
  factors, 8, 16, 28, 40, 92
  status, 14
  traits, 23
Emotionally Unstable, 8, 22
Encoding, 86
Endogenous, 19, 27, 82
Enrichment, 85
Environment, 82
  arousing curiosity, 14
  exploration of, 12, 25
  features of schools used in study,
    41-42
  impoverished, ix
  individual &, 30
Epilepsy, 19
Evaluation, 5
  of research, 79, 84
  of test results, 43
  of this study, 36
Exogenous, 19, 27, 82
Expectancy Table, 63
Experience, 86, 90
  appropriate, 17
  challenging, 14
  common learning, 53
  cultural and educational, 72-73
  difference in intelligence, 4
  difficulties in arithmetic, 6
  diversities in development, 6
  impoverished, ix, 23
  length of, 71
  number concepts, 12
  ordinary, 19
  perceptual, 73, 75
  pre-school, 11-12
  physical, 12, 13, 15
  prior, 14
  sensorimotor, 25, 71
Explore
  environment, 25
Extraneous
  facts, 7
  material, 20
  stimulation, 34

stimuli, 28
Extrinsic, 31
Eye-Motor Co-ordination, 74, 87
Eye-Movements, 30
  horizontal saccadic & pursuit, 25
  & perceptual integration, 24

Fading, 32
Feedback, 57
Feeding, 29
Figure
  & background, 19, 28, 30
  -ground, 74
  -ground discrimination, 87
  -ground perceptual ability, 20, 80
Finger Dexterity, 44
Finger Localisation, 11
Finger Schema, 11
Form Constancy, 74, 87
Formal Instruction, 14, 17
Forms, 24
Four Basic Rules, x
Frostig Test, 74, 87, 95
Full-Term, 29

Generalisation, 21, 30
Gerstmann Syndrome, 10
Gestalt, 18
  maturation, 26
  appreciation, 26
Girls
  & arithmetic attainment, 9-10
  premature children, 29
Goodenough Draw-A-Man Test, 43, 93

Hearing, 43
  loss, 22
Hemiplegia, xi
  double, xi
  left & right, 20
Hyperactive, 28, 83
  hyperactivity, 73
Hypotheses, 35, 36
  hypothesis, 71

I.Q., 83, 85-86, 103
  intercorrelations of, 71, 75
  predictive value of, 85

& severity of handicap, 23
Illinois Psycholinguistic Battery, 86
Impoverished, 23
Inattentive, 28
Incentives, 66, 82, 88, 89
Incidence, 18
  of arithmetic difficulties, 79
  of behaviour disturbance, 28
  of emotional disturbance, 30
  of ocular defects, 29
  sex, 49
Infants
  learning in, 24
  schools, 19
Infection, 18
Injury to the Brain, 18, 25, 25-26,
    26-27, 70-75, 82
  & arithmetic attainment, 26-27
  congenital, 25-26
Integrate, 21
  integrated, 79
  integrated approach, 79, 90
  integrated movements, 28
  integrated in purpose, 90
  integrated relationship, 90
  integrating, 19
  integrating constituent elements, 30
  integration, 24, 51, 104
  perceptual integration, 24, 80
Integrity, 24
Intellectual
  abilities, ix, 9, 92
  abnormalities, 19
  attributes, 39
  background, 12
  capacity, 35
  data, 93
  demands, 94
  development, 25, 74
  disorders, 17
  experiences, 6
  factors, 8, 40
  functioning, 85, 86
  growth, 86
  loss, 23
  task, 73
Intelligence
  & ability to sit, 22

& arithmetic attainment, 6, 16, 70-73
differences in, 23
& learning ability, 6
level of, 94
levels of, in non c.p. children, 42
& motor disorders, 22
& new experiences, 6
& P.I., x
"pure," 72, 73
& success in arithmetic, 16
variations in, 23
Intelligence Tests, 23-24, 49, 50-52, 68, 103
tests of intelligence, 73, 86
tests of "pure" intelligence, 72
Interpersonal
contacts, 57
difficulties, 94
relationships, 89, 90
Intrinsic, 32
Intuitive, 14, 81
Invariant, 14

Language, 86
development, 7, 16, 21
& P.I., 33
processes, 22
Learning, 43, 82, 83, 93
ability & differences in intelligence, 6
animal studies, 81
arithmetic, 73, 92
"blocks," 10
in the classroom, 81
common experience, 53
confidence in, 44
children's, 81
difficulties, 5, 84
difficulties of c.p. children, 17-23
effects on, 39
in infants, 24
mathematical—"operational character" of, 12
& ocular disorders, 24
pace of, 36
past, 72
programmed, 87
& P.I., ix
problems, 84
process of, 73

rates, 14
readiness, 84
reinforcement of, x, 31-32
in schools, 23
situation, 14
theory, 80, 82
Lesions
in c.p. & impairment in arithmetic attainment, 26
within the CNS, 35
Linear, 31, 32, 34, 43, 56
Logical Number, 14
Longitudinal, 81

Maladjustment, 104
& fear of arithmetic, 9
& reading, 8
Manipulate, 25, 73
teaching machine, 93, 94
manipulation of concrete materials, 12
manipulation of play materials, 12
Manipulative skills, ix
Matching, 24
difficulties in, 30
procedure, 39
rods, 12
Matrix, 60-63
Maturation
Gestalt, 26
variations in, 6
"visual gestalt maturational level," 68
Measurement
of arithmetic attainment, x, 93
of intellectual & motor factors, 43
Measures, 24
of arithmetic attainment, 10, 46-47
Mechanical
aids, 41
methods of teaching arithmetic, 13
skills, 13, 14
Memory, 72
abilities, 23
processes, 22
short term, 73
verbal, 26
visual, 26, 68

Mental
 age, 83
 age norms, 68
 arithmetic, 47
 arithmetic problems, 69
 capacity, 22
 "development," 21
 operations, 24
Mental Defectives
 institutionalized female, 27
 mental deficiency & c.p., xi
Mental Operations, 24
Mentally Retarded, xi
 & arithmetic attainment, 6-7, 16
 & Finger Schema, 11
 & neurological factors, 10
 & P.I., 33, 34
Minimal Brain Damaged, 83
Minimal Cerebral Dysfunctioning, 83
Mobility, ix
Modified, 30
Motivation, 5
 of parents, 12
Motor
 defects, 24
 development, 83
 disorders, 17, 22, 25-26
 encoding, 86
 handicaps, 35, 70
 impairments, 83
Multiple Choice, 32
Multiple Handicaps, 36
Multiplying, 8
 multiplication, 15, 45, 47, 104

Neurological
 abnormalities, 80
 examination, 35
 experiences & arithmetic, 6
 factors & arithmetic, 10-12
 impairment, 28, 83
Neuromotor, 18
Non-Academic, 58, 59
Normal Children, xi, 11, 19, 29, 42, 43, 70, 74, 83, 94
 & arithmetic, ix, 5, 6, 16, 45
 & brain-injured, 21
 & c.p. children, 18-20, 23, 24-25, 30
 & emotional factors, 28

& incentives, 66
& learning, ix, 81
& mentally retarded, 7, 16
& P.I., 32
& reinforcement techniques, 81
& teaching machines, 65
Novelty Value, 66
Number Combinations, 15
Number Concepts, 73, 81
 development of, 24
 formation of, 12, 27
 & pre-school experiences, 17
Number Scheme, 73
Number System, 27
Number Vocabulary
 formation of, 12
 & pre-school experiences, 17

Objects, 74
 discovery of properties of, 12
 forming ideas of, 12
 handling of, 28
 manipulate, 25
 manipulation of, 17
 sorting, 19
 in space, 25, 73
Ocular Defects, 24
Odd-Even Reliability, 61
Operant Conditioning, 31, 80
Operational
 "character," 12
 stage, 14
Operations
 bodily, 12
 mode of, 7
 significance of, 12
Optimal Conditions, 82
Overprotection, 28

Pace, 58, 86
 of learning, 36, 56
Pathological, 80, 83
Perceptual
 ability, 20, 73
 approximations, 14
 components, 72
 defects, 25
 development, 21
 difficulties, 73, 81

Index

discrimination, 20
disorders, 17, 20, 27, 30, 35, 83, 92
disturbances, 19
experiences, 73, 75
functioning, 74
impairment, ix, 20
integration, 24, 80
processes, 22
responses, 17
scheme of visual spatial organisation, 27
skills, 17, 23
tasks, 18, 20, 30
weaknesses, 24
Perinatal, 29
Peripheral
  abilities, 23
  disorders, 24, 25
Perseveration, 51, 61, 62, 73, 80, 104
  perseverate, 19, 30
Personality
  development, 29
  disorders, 29
  ratings, 21
  studies of, 29
Physical
  actions, 12
  care, 41
  experiences, 12, 13, 15
  handicap, 35
Physically Handicapped, 24
  & c.p., 25-26
  schools for, 35
  & teaching machines, 65
Pigeons, 31
Play
  activities, 25
  materials, 12
  situations, 12
Practice, 59
  at home, 16
  computational, 15, 17
  teaching, 18
Preconceptual, 81
Predictive Validity, 23, 86
Predictor Variables, 49, 50
Prematurity, 28
Prenatal, 29
Preoperational, 14

Preschool
  arithmetic achievement, 12, 17
  experiences, 6, 11-12, 23, 25
  period, 29
Primary
  grade children in America, 12
  grade teachers, 13
  school, 11
"Primary Arithmetic" Programme, 44-45, 93
Procedure, 39
  mechanical, 13
Programmed Instruction, ix-xi, 39
  anxieties about, 41-42
  & arithmetic, 5, 30-34, 95
  & c.p. children, 5, 34-36, 65-68
  & Skinner, ix
  teachers' comments on, 56-60
  & teaching machines, 83-85
Prompts, 31
Prospective, 29
Psychiatrically At Risk, 28
Psycho-analytic, 10
Punishment, x, 31
Pupil Satisfaction, 36, 57-58
  & teacher response, 66-67
Pursuit, 25

Qualitative
  & quantitative arithmetic work habits, 27
  qualitatively better results—teaching machine, 68
Quantitative
  "expression" of young normal children, 5
  & qualitative arithmetic work habits, 27
  relationships, 14
Quartiles, 63

Rapport, 14
Ratios, 31
Reactions
  of children & teachers to programme, 46
  of children & teachers to P.I., 57-60
  compensatory, 35

Reading, 19, 31, 34
  ability, 33
  & anxious children, 8
  & arithmetic, 26
  attainment in c.p. children, 22, 23, 26
  backwardness, 11
  & branching programme, 44
  learning, 73
  & maladjustment, 8
  practice at home, 16
  teaching of, 19
  weaknesses in, 20
Reasoning
  ability, 7, 15, 23
  & assessment of arithmetic attainment, xi
  mathematical, 68
  processes, 22
  verbal, 26, 68
  visual, 26
Recall, 33
Recognise, 10, 24
  recognising, 74
Reinforcement, x, 32, 44, 89
  reinforcing, 30, 31, 89
  reinforced, 89
  technique, 81
Rejection, 28
Remedial
  education, 20
  exercises, 81
  programme, 87
  purposes, 85
  teaching, 9
  techniques, 20
Repertoires, 32
Research, ix
  findings reviewed, 5
  needs & methods, 91
  strategy, 6
Responses, 19, 30, 31, 34
Restlessness, 28, 73, 104
Retrospective, 28
Right-Left Discrimination, 11
Rigidity, 19
Rote
  counting, 12
  learning, 33

Saccadic, 25
Schonell Test, 46, 47, 53, 61, 62, 69, 70, 71, 72, 93, 104
School Experiences, 12-16
  impoverished, 23
  source of difficulty, 6
Selective, 23
Sensory
  channels, 25
  defects in c.p. children, 22, 80
  disorders, 43, 92
  factors, 43
Sensori-Motor, 17
  activities, 12
  experiences, 25, 29, 71
Sequence, 73, 81
  of concept development, 14
  logical, 19
  systematic, 33
Sequential
  maturation, 6
  ordering (of the fingers), 80
  stages, 6
  steps, 31
Shapes, 24
Skinner, ix, x, 30-32, 81, 82, 95
Skinnerian, 31
Sociological, 79
Southend Test, 26, 46-47, 53, 70, 71, 72, 93, 104
Space, 74
  knowledge of objects in, 25
  perception, 83
  position in, 74, 80
Spastic, xi, 42, 44, 50
  & athetoids, 20, 23, 26
  & normals, 18
  & this study, xii
Spatial
  ability, 24, 87
  components, 72
  concepts, 71
  difficulties, 74-75, 87, 94
  disorders, 73-75, 83
  functioning, 74
  "inability," 75
  knowledge, 12, 17
  organisation, 27

Index

relations, 74, 78
& WISC, 50
Specific Learning Disability, 17
Speech, 83
  backwardness in, 82
  & c.p. children, 22
  difficulties in, 18
Squint, 24, 29
  squinting, 25
Stable, 29
Stanford Binet, 21, 24, 27, 68, 70, 72, 73, 83
Statistical Analysis, 48
Strabismus, 24
Structural Apparatus, 14, 17
Subjects, 42
Subnormals, 81
Subtracting, 8
  subtraction, 15, 26, 47, 104
Syllabus, 45
Symbols, 25
Synthesised, 74

Tactual-Motor
  performance, 28
  skills, 80
Teacher, 45, 88, 89, 93, 95
  abilities of, 14, 17
  & administrative tests, 43
  anxieties of, 41-42
  attitude of, 10, 12-13, 17, 88
  awareness of, 16, 17
  children's likes for, 39
  control & supervision of class, 42
  implications for, 79-91
  & new methods, 87
  & operant conditioning, 31
  personality of, 45
  of the physically handicapped, 95
  primary grade, 13
  & P.I., 11 12, 56-60
  quality of instruction, 39
  reactions to programme, 46
  & research, 89-90
  response, 66-67
  role of, ix, 12-13, 14, 15, 88
  satisfaction, 58
  understanding of, 5, 12-13, 17

Teaching, 60
  aids, 14-15, 41
  appropriate intervention in, 13-14
  arithmetic, x, 12, 14, 32-33, 80, 87
  of c.p. children, 80, 82
  conventional, 34-35, 36, 55, 65, 68, 94
  facilities, 40
  formal, 42
  mathematical, 28
  of mentally retarded brain-injured children, 19
  methods, 31, 33, 39, 53, 84, 85, 86, 87, 88
  & motor disorders, 22
  & new techniques, 87-91
  planning, 87
  practice, 35
  programme, 45-46, 52, 82
  styles, 45, 95
  systematic, 15
  technology of, 83
  unskilled, 9
  unsympathetic, 9
Teaching Machine, 32-33, 40, 42, 43-46, 66, 67-68, 88, 89, 93, 94
  & P.I., 83-85
  in schools for normal children, 65
  teachers' comments on, 56-60
Technological, 31
Temperamental Characteristics
  & arithmetic ability, 28
  & arithmetic disability, 8, 16
Terminal, 30
Terminology, 82
Testing, 32
Therapist, 90
  therapy staff, 41
Thinking, 81, 88
  & concept development, 5
  disturbances in, 18
  educational, 31
Thought, 82
Traits, 23
Transfer, 33
  ability, 33, 82
  of learning, 5

Unsettledness, 8

Variables, 35, 48-57, 60, 103-104
 effects of, 39
 personality, 73
Variable Intervals, 31
Verbal
 abilities, 22
 behaviour, 12
 instruction & deaf, 60
 memory, 26
 & non verbal IQs, 21
 reasoning ability, xi, 26, 68
 WISC scores, 11, 50
Visual
 disorders in c.p. children, 24
 functioning, 24
 "gestalt maturational level," 68
 impairments, 22
 memory, 26, 68
 perception, 24
 perceptual disorders, 20
 perceptual skills, 17
 perceptual tasks, 18
 reasoning, 26
 spatial organisation, 27
 stimuli, 19, 86
 tasks, 24
Visuomotor
 association, 86
 disorders, 24, 27, 83
 disturbance, 25
 experiences, 75
 functioning, 74
 impairment, ix, 25, 81
 perceptual tasks, 18
 sequencing, 86
 skills, 23, 25, 30, 80
 tasks, 20, 28, 30
Vocabulary, 26

Wheelchairs, 44
WISC, xi, 43, 50, 51, 61, 62, 68, 69, 71, 72, 73, 74, 86, 93, 103
Writing, 19, 34
 backwardness, 11
 & Gerstmann Syndrome, 10
 learning to write, 73
 skills, 20
 teaching of, 19